INTRODUCTION TO
QUANTUM
MECHANICS

Solutions to Problems

INTRODUCTION TO
QUANTUM
MECHANICS
Solutions to Problems

John Dirk Walecka
College of William and Mary, USA

World Scientific

EW JERSEY · LONDON · SINGAPORE · BEIJING · SHANGHAI · HONG KONG · TAIPEI · CHENNAI · TOKYO

Published by

World Scientific Publishing Co. Pte. Ltd.
5 Toh Tuck Link, Singapore 596224
USA office: 27 Warren Street, Suite 401-402, Hackensack, NJ 07601
UK office: 57 Shelton Street, Covent Garden, London WC2H 9HE

Library of Congress Control Number: 2021037014

British Library Cataloguing-in-Publication Data
A catalogue record for this book is available from the British Library.

INTRODUCTION TO QUANTUM MECHANICS
Solutions to Problems

ISBN 978-981-124-464-3 (hardcover)
ISBN 978-981-124-525-1 (paperback)
ISBN 978-981-124-465-0 (ebook for institutions)
ISBN 978-981-124-466-7 (ebook for individuals)

For any available supplementary material, please visit
https://www.worldscientific.com/worldscibooks/10.1142/12493#t=suppl

For John and Ann

Preface

The author recently published a book entitled *Introduction to Electricity and Magnetism* [Walecka-18]. It is based on an introductory course taught several years ago at Stanford, with over 400 students enrolled. The only requirements were an elementary knowledge of calculus and familiarity with vectors and Newton's laws; the development was otherwise self-contained. The lectures, although relatively concise, take one from Coulomb's law to Maxwell's equations and special relativity in a lucid and logical fashion.

Although never presented in an actual course, it occurred to the author that it would be fun to compose an equivalent set of lectures that would serve as a *prequel* to that *Electricity and Magnetism* text. This book has now also been published as *Introduction to Classical Mechanics* [Walecka-20]. The goal of this second text is to provide a clear and concise set of lectures that take one from the introduction and application of Newton's laws up to Hamilton's principle and the lagrangian mechanics of continuous systems.

Both of these texts on classical physics are meant for initial one-quarter physics courses. These lectures, aimed at the very best students, assume a good concurrent course in calculus; they are otherwise self-contained. Both texts contain an extensive set of accessible problems that enhances and extends the coverage. As an aid to teaching and learning, the solutions to these problems have now been published in additional texts [Walecka-19, Walecka-21].

A third just-published text completes the first-year introduction to physics with a set of lectures on *Introduction to Quantum Mechanics*, the very successful theory of the microscopic world [Walecka (2021a)]. The Schrödinger equation is motivated and presented. Several applications are explored, including scattering and transition rates. The applications are extended to include both quantum electrodynamics and quantum statistics. There is a discussion of quantum measurements. The lectures then arrive at a formal presentation of quantum theory together with a summary of its postulates. A concluding chapter provides a brief introduction to relativistic quantum mechanics. An extensive set of accessible problems again enhances and extends the coverage. The current book provides the solutions to those problems.

The goal of these three texts is to provide a good, understandable, one-year introduction to the fundamentals of classical and quantum physics. It is my hope that students and teachers alike will find the use of these books rewarding and share some of the pleasure I took in writing them.

Quantum mechanics is a huge field, and no attempt has been made to provide a complete bibliography. The references given are only directly relevant to the discussion at hand. It is important, however, to mention some of the good, existing books on quantum mechanics that the author has found particularly useful, such as [Wentzel (1949); Bjorken and Drell (1964); Bjorken and Drell (1965); Schiff (1968); Itzykson and Zuber (1980); Landau and Lifshitz (1981); Shankar (1994); Merzbacher (1997); Gottfried and Yan (2004); Feynman and Hibbs (2010)]. In addition, Appendix B lists some significant names in quantum mechanics, both in its theory and in its applications.

I would like to once again thank my editor, Ms. Lakshmi Narayanan, for her help and support on this project. I am also grateful to Paolo Amore for his reading of the manuscript of the quantum mechanics text.

Williamsburg, Virginia   *John Dirk Walecka*
May 17, 2021   *Governor's Distinguished CEBAF*
                                            Professor of Physics, Emeritus
                                            College of William and Mary

Contents

Chapter 1

Motivation

Problem 1.1 The discussion started with a classical wave that is the real part of

$$\Psi(x,t) = e^{i(kx-\omega t)}$$

Suppose we have two oscillating functions with the same frequency

$$\text{Re}\left(\alpha e^{-i\omega t}\right) = \frac{1}{2}\left(\alpha e^{-i\omega t} + \alpha^* e^{i\omega t}\right)$$

$$\text{Re}\left(\beta e^{-i\omega t}\right) = \frac{1}{2}\left(\beta e^{-i\omega t} + \beta^* e^{i\omega t}\right)$$

and suppose we are interested in the *product* of these functions, as in the energy, energy flux, etc. Furthermore, suppose we want the *time average* of such products, which we denote by $\langle \cdots \rangle$.

(a) Show

$$\left\langle \text{Re}\left(\alpha e^{-i\omega t}\right) \text{Re}\left(\beta e^{-i\omega t}\right) \right\rangle = \frac{1}{4}\left(\alpha\beta^* + \alpha^*\beta\right) = \frac{1}{2}\text{Re}\left(\alpha\beta^*\right)$$

(b) Suppose the wave function is a superposition of two such terms

$$\Psi(x,t) = \text{Re}\left(\alpha e^{-i\omega t}\right) + \text{Re}\left(\beta e^{-i\omega t}\right)$$

Show

$$\left\langle \Psi^2(x,t) \right\rangle = \frac{1}{2}|\alpha + \beta|^2$$

These are extremely useful relations when using complex solutions to the classical wave equation while describing real situations.[1]

---

[1]See [Fetter and Walecka (2003)].

Solution to Problem 1.1 We define the time average as

$$\langle \cdots \rangle = \mathrm{Lim}_{T \to \infty} \frac{1}{T} \int_0^T dt \cdots$$

It follows that

$$\langle e^{i\omega t} \rangle = \langle \cos \omega t + i \sin \omega t \rangle = 0$$

while the time average of a constant is just that constant.

(a) Consider

$$\langle \mathrm{Re} \left(\alpha e^{-i\omega t} \right) \mathrm{Re} \left(\beta e^{-i\omega t} \right) \rangle = \frac{1}{4} \langle \left(\alpha e^{-i\omega t} + \alpha^\star e^{i\omega t} \right) \left(\beta e^{-i\omega t} + \beta^\star e^{i\omega t} \right) \rangle$$

$$= \frac{1}{4} (\alpha^\star \beta + \alpha \beta^\star) = \frac{1}{2} \mathrm{Re} \left(\alpha \beta^\star \right)$$

(b) If the wave function is a superposition of two such terms, then

$$\langle \Psi^2(x, t) \rangle = \langle \left[\mathrm{Re} \left(\alpha e^{-i\omega t} \right) + \mathrm{Re} \left(\beta e^{-i\omega t} \right) \right] \left[\mathrm{Re} \left(\alpha e^{-i\omega t} \right) + \mathrm{Re} \left(\beta e^{-i\omega t} \right) \right] \rangle$$

$$= \frac{1}{4} \langle \left[\left(\alpha e^{-i\omega t} + \alpha^\star e^{i\omega t} \right) + \left(\beta e^{-i\omega t} + \beta^\star e^{i\omega t} \right) \right] \times$$

$$\left[\left(\alpha e^{-i\omega t} + \alpha^\star e^{i\omega t} \right) + \left(\beta e^{-i\omega t} + \beta^\star e^{i\omega t} \right) \right] \rangle$$

$$= \frac{1}{2} \left[|\alpha|^2 + |\beta|^2 + (\beta \alpha^\star + \beta^\star \alpha) \right] = \frac{1}{2} |\alpha + \beta|^2$$

As stated, these are extremely useful relations when using complex solutions to the classical wave equation while describing real situations.

Problem 1.2 Consider a classical plane wave incident on two slits separated by a distance d. If we seek the amplitude of the wave on the other side of the slits, we can just add the waves $\mathrm{Re}[A e^{i(kx - \omega t)}]$ coming from each slit, where x is the distance to the observing screen. For small angles, the difference in optical pathlength Δ from the second slit is

$$\Delta \approx \theta d$$

where d is the slit separation (see Fig. 1.1 below).

(a) Use the result in the previous problem to show that the time-average intensity pattern on the observing screen is proportional to

$$\langle \Psi^2(x, y) \rangle = 2|A|^2 \cos^2 \left(\frac{k\Delta}{2} \right) \qquad ; \text{ interference pattern}$$

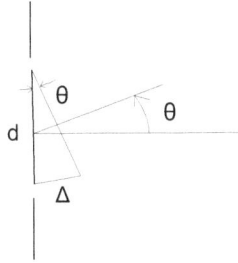

Fig. 1.1   Difference in optical pathlength Δ for two-slit interference pattern measured at an angle θ. A plane wave is incident from the left, and the interference pattern shows up on a screen to the right. The transmitted wave is uniform in the direction normal to this plane.

(b) Show the condition for the first interference minimum is then

$$\theta \approx \frac{\lambda}{2d} \qquad ; \text{ first minimum}$$

Note that no matter how small the slit separation d, if the incident wavelength λ is comparable to it, then this interference pattern will show up at a *finite angle* on the screen.

Solution to Problem 1.2 We have a classical plane wave incident from the left in Fig. 1.1 above, which uniformly covers the two slits. Each slit then acts as a source that emits a wave that strikes the observing screen on the right, the observing screen is parallel to the slit screen and a distance x from it. From Huygen's principle, we can just add the two waves at the observing screen, and the wave at the screen is given by

$$\Psi(x,t) = \mathrm{Re}[Ae^{i(kx_1 - \omega t)}] + \mathrm{Re}[Ae^{i(kx_2 - \omega t)}]$$

where (x_1, x_2) are the actual distances from the slit to the point of observation.[2] For small angles, the difference in optical pathlength Δ from the second slit is

$$\Delta \approx \theta d$$

where d is the slit separation (see Fig. 1.1 above).

───────────────

[2]The outgoing waves are actually cylindrical waves, and we have incorporated an average value of the distance dependence of the amplitude of such waves into A.

(a) It then follows from the previous problem that

$$\langle \Psi^2(x,t) \rangle = \frac{1}{2}|A|^2 |e^{ikx_1} + e^{ik(x_1+\Delta)}|^2 = \frac{1}{2}|A|^2 \left|1 + e^{ik\Delta}\right|^2$$

$$= = \frac{1}{2}|A|^2 \left|e^{ik\Delta/2} + e^{-ik\Delta/2}\right|^2 = 2|A|^2 \cos^2\left(\frac{k\Delta}{2}\right)$$

(b) On the axis, the time-average intensity is just $2|A|^2$. This intensity then vanishes at a series of angles θ off-axis, where the first minimum occurs at $k\Delta = \pi$. This is rewritten as

$$\theta \approx \frac{\lambda}{2d} \qquad\qquad ; \text{ first minimum}$$

From Fig. 1.1, this condition is satisfied when the wave from the second slit is just out of phase with the wave from the first slit, having traveled an additional distance $\Delta = \lambda/2$.

Problem 1.3 (a) Suppose we have a physical system satisfying the classical wave equation and obeying periodic boundary conditions (p.b.c.) in one dimension

$$k = \frac{2\pi n}{L} \qquad\qquad ; n = 0, \pm 1, \pm 2, \cdots$$

Write the differential form of this relation as

$$dk = \frac{2\pi}{L} dn$$

Use $\omega = kc$ to conclude that the number of normal modes per unit length is given by

$$\frac{1}{L} dn = \frac{1}{c} d\nu$$

(b) Extend this result to two dimensions, and show

$$\frac{1}{L^2} d^2 n = \frac{1}{L^2} dn_x dn_y = \frac{1}{c^2} d\nu_x d\nu_y$$

If this is summed over all modes with a given length ν, then show the number of modes per unit area is

$$\frac{1}{L^2} d^2 n = \frac{2\pi}{c^2} \nu \, d\nu$$

(c) Show that in three dimensions, the number of modes per unit volume of frequency ν is given by

$$\frac{1}{L^3} d^3 n = \frac{4\pi}{c^3} \nu^2 \, d\nu$$

If there is a degeneracy of g for the types of normal modes at a given frequency, then

$$\frac{1}{L^3} d^3 n = \frac{4\pi g}{c^3} \nu^2 \, d\nu$$

(d) With the use of $g = 2$ for the two transverse polarizations of a free electromagnetic wave, and with the equipartition result for the energy of a normal mode in Eq. (1.4), show that one obtains the following classical expression for the electromagnetic energy density in a cavity

$$U(\nu, T) = \frac{8\pi k_B T \nu^2}{c^3} \qquad ; \text{ classical energy density}$$

Solution to Problem 1.3 (a) Consider a physical system satisfying the classical wave equation and obeying periodic boundary conditions (p.b.c.) in one dimension. In this case, the wavenumbers k are given by[3]

$$k = \frac{2\pi n}{L} \qquad ; n = 0, \pm 1, \pm 2, \cdots$$

Since the modes are very closely spaced as the interval becomes very large $(L \to \infty)$, one can group them together in the small interval dk and write

$$dn = \frac{L}{2\pi} dk$$

This is a very important relation that allows us to convert a sum to an integral.

If we employ the relation between wavenumber and frequency, $k = 2\pi/\lambda = 2\pi\nu/c$, then the number of modes per unit length is

$$\frac{1}{L} dn = \frac{1}{c} d\nu$$

(b) In two dimensions, the number of modes is given by

$$d^2 n = \left(\frac{L}{2\pi}\right)^2 d^2 k$$

---

[3] Another visualization of p.b.c. is a linear chain of segments of length L, with identical physics imposed on each segment.

In terms of frequency, this is

$$\frac{1}{L^2}d^2n = \frac{1}{L^2}dn_x dn_y = \frac{1}{c^2}d\nu_x d\nu_y$$

If this is summed over all modes with a given length ν, then the number of modes per unit area in two dimensions is

$$\frac{1}{L^2}d^2n = \frac{2\pi}{c^2}\nu\, d\nu$$

(c) This result is immediately extended to three dimensions, with

$$d^3n = \left(\frac{L}{2\pi}\right)^3 d^3k$$

If there is a degeneracy of g for the types of normal modes at a given frequency, then

$$d^3n = g\left(\frac{L}{2\pi}\right)^3 d^3k$$

The number of modes per unit volume of frequency ν follows as

$$\frac{1}{L^3}d^3n = \frac{4\pi g}{c^3}\nu^2\, d\nu$$

(d) With the use of $g = 2$ for the two transverse polarizations of a free electromagnetic wave, and with the equipartition result for the energy of a normal mode in Eq. (1.4), one obtains the following classical expression for the electromagnetic energy density in a cavity

$$U(\nu, T) = \frac{8\pi k_B T\nu^2}{c^3} \qquad ;\ \text{classical energy density}$$

Since there is no limit on how small the wavelength is here, or on how high the frequency, this energy density grows without bound as ν^2. The result is the classical *ultraviolet catastrophe* for the energy density in a cavity.

Problem 1.4 A photon has a frequency just into the ultraviolet

$$\nu = 10^{15}\ \text{Hz}$$

What is its wavelength? What is its energy? What is its momentum?

Solution to Problem 1.4 We need two constants in SI units

$$c = 2.998 \times 10^8\ \text{m/s}$$
$$h = 6.626 \times 10^{-34}\ \text{Js}$$

The wavelength and energy of the photon are then given by

$$\lambda = \frac{c}{\nu} = \frac{2.998 \times 10^8}{10^{15}} \, \text{m} = 2.998 \times 10^{-7} \, \text{m}$$
$$E = h\nu = 6.626 \times 10^{-34} \times 10^{15} \, \text{J} = 6.626 \times 10^{-19} \, \text{J}$$

The momentum of the photon is given by[4]

$$p = \frac{E}{c} = \frac{6.626 \times 10^{-19}}{2.998 \times 10^8} \, \text{Ns} = 2.210 \times 10^{-27} \, \text{Ns}$$

Problem 1.5 (a) In the books on *Introduction to Classical Mechanics* it is shown that the energy E and radius a of a particle of mass μ and charge e performing circular orbits about a point charge $-Ze$ are

$$E = -\frac{Ze^2}{4\pi\varepsilon_0} \frac{1}{2a} \qquad ; \text{energy}$$
$$a = \frac{4\pi\varepsilon_0}{Ze^2} \frac{\vec{L}^2}{\mu} \qquad ; \text{radius}$$

where $|\vec{L}| = \mu a v$ is the angular momentum. Verify these results;

(b) We shall show at the beginning of the next chapter that de Broglie's relation for the wavelength of a particle implies that the angular momentum of the above system is *quantized* as $|\vec{L}| = n\hbar$ with $\hbar = h/2\pi$ where h is Planck's constant, and n is an integer $n = 1, 2, \cdots$. Show this immediately yields Bohr's quantum theory of the spectrum of one-electron atoms.[5]

Solution to Problem 1.5 (a) For circular motion with radius a in the Coulomb potential, Newton's second law gives

$$\frac{\mu v^2}{a} = \frac{Ze^2}{4\pi\varepsilon_0 a^2}$$

The energy is then

$$E = \frac{1}{2}\mu v^2 - \frac{Ze^2}{4\pi\varepsilon_0 a} = -\frac{Ze^2}{4\pi\varepsilon_0} \frac{1}{2a}$$

The square of the angular momentum is

$$\vec{L}^2 = (\mu a v)^2 = \mu a \frac{Ze^2}{4\pi\varepsilon_0}$$

---

[4] Note that $1\,\text{J} = 1\,\text{kg(m/s)}^2$, and $1\,\text{Ns} = 1\,\text{kg(m/s)}$.
[5] See [Walecka (2008)].

Hence the radius of the orbit is given by

$$a = \frac{4\pi\varepsilon_0}{Ze^2} \frac{\vec{L}^2}{\mu}$$

(b) We give an argument at the start of Chapter 2 that the angular momentum is quantized according to[6]

$$|\vec{L}| = n\hbar \qquad ; n = 1, 2, 3, \cdots$$

This immediately yields Bohr's theory of the atom. For an electron performing a circular orbit of radius a about a point nuclear charge $-Ze$, one has[7]

$$a = \frac{n^2}{Z\alpha} \frac{\hbar}{m_e c} \qquad ; n = 1, 2, 3, \cdots$$

$$E = -\frac{(Z\alpha)^2}{2n^2} m_e c^2$$

Here α is the fine-structure constant, $\hbar/m_e c$ is the electron Compton wavelength, and $m_e c^2$ is its rest mass

$$\alpha = \frac{e^2}{4\pi\varepsilon_0 \hbar c} = \frac{1}{137.0}$$

$$\frac{\hbar}{m_e c} = 3.862 \times 10^{-13}\,\mathrm{m}$$

$$m_e c^2 = 0.5110\,\mathrm{MeV}$$

Problem 1.6 (a) Suppose we analyze a photon-electron collision through the conservation of momentum and energy

$$\vec{p}_e = \vec{p}_0 - \vec{p}_1 \qquad ; \text{momentum conservation}$$

$$h\nu_0 = h\nu_1 + \frac{\vec{p}_e^2}{2m} \qquad ; \text{energy conservation}$$

---

[6]See also Prob. 2.2.

[7]We include Bohr's theory of the atom with its circular orbits as *motivation*. In quantum mechanics, the Schrödinger equation must be solved in a Coulomb potential, as done, for example, in [Walecka (2013)]. Interestingly enough, one then obtains the same energy spectrum.

Show the substitution of the first relation in the second leads to

$$h(\nu_0 - \nu_1) = \frac{1}{2m}(\vec{p}_0 - \vec{p}_1)^2 = \frac{1}{2m}\left(\frac{h}{c}\right)^2 (\nu_0^2 + \nu_1^2 - 2\nu_0\nu_1\cos\theta)$$

$$\nu_0 - \nu_1 = \frac{h}{mc^2}\nu_0\nu_1\left[(1 - \cos\theta) + \frac{(\nu_0 - \nu_1)^2}{2\nu_0\nu_1}\right]$$

(b) Show that if the energy shift is small, the last term can be neglected for all θ of interest

$$\frac{(\nu_0 - \nu_1)^2}{2\nu_0\nu_1} \ll 1$$

(c) The frequency of the light is related to its wavelength by $\nu_0 = c/\lambda_0$ and $\nu_1 = c/\lambda_1$. Hence show that one arrives at the lovely, simple Compton formula for the shift in wavelength[8]

$$\lambda_1 - \lambda_0 = \frac{h}{mc}(1 - \cos\theta) \qquad ; \text{ Compton formula}$$

This is in complete agreement with the experimental results. The results on Compton scattering confirmed the particle nature of the photon introduced by Einstein in his explanation of the photoelectric effect.

Solution to Problem 1.6 (a) If a photon of energy $h\nu$ and momentum $h\nu/c$ is scattered from a target particle of mass m, originally at rest, the conservation of momentum and energy read as follows

$$\vec{p}_e = \vec{p}_0 - \vec{p}_1 \qquad ; \text{ momentum conservation}$$

$$h\nu_0 = h\nu_1 + \frac{\vec{p}_e^2}{2m} \qquad ; \text{ energy conservation}$$

where $(\vec{p}_0, \vec{p}_1)$ are the initial and final momenta of the photon, and $\vec{p}_e$ is the recoil momentum of the target particle.

Substitution of the first relation in the second gives

$$h(\nu_0 - \nu_1) = \frac{1}{2m}(\vec{p}_0 - \vec{p}_1)^2$$
$$= \frac{1}{2m}\left(\frac{h}{c}\right)^2 (\nu_0^2 + \nu_1^2 - 2\nu_0\nu_1\cos\theta)$$

---

[8]With the use of proper relativistic kinematics for the particle, this result holds without approximation.

where θ is the scattering angle. This is rewritten as

$$\nu_0 - \nu_1 = \frac{h}{mc^2}\nu_0\nu_1 \left[(1 - \cos\theta) + \frac{(\nu_0 - \nu_1)^2}{2\nu_0\nu_1}\right]$$

(b) For a small energy shift, the last term can be neglected for all θ of interest, an approximation that can always be verified at the end,

$$\frac{(\nu_0 - \nu_1)^2}{2\nu_0\nu_1} \ll 1$$

(c) The frequency of the light is related to its wavelength by $\nu_0 = c/\lambda_0$ and $\nu_1 = c/\lambda_1$. Hence one arrives at the Compton formula for the shift in wavelength

$$\lambda_1 - \lambda_0 = \frac{h}{mc}(1 - \cos\theta) \qquad ; \text{ Compton formula}$$

As stated, this is in complete agreement with the experimental data, and the results on Compton scattering confirmed the particle nature of the photon introduced by Einstein in his explanation of the photoelectric effect.

Chapter 2

Wave Packet for Free Particle

Problem 2.1 Suppose one repeatedly prepares a non-relativistic particle in the plane-wave state of definite momentum in Eq. (2.5) and sends it against the two-slit opening in Fig. 12.1 in the text. Show the measured particle density on the screen will exhibit an interference pattern *identical* to that in Prob. 1.2.

Solution to Problem 2.1 The plane-wave state of definite momentum in Eq. (2.5) is

$$\Psi(x,t) = e^{i[kx - \omega(k)t]} = e^{i[kx - (\hbar k^2/2m)t]}$$

Since this is a solution to the Schrödinger wave equation, we can just superpose the waves from the two slits, exactly as in Prob. 1.2. The wave at the observing screen is thus given by

$$\Psi(x,t) = Ae^{i(kx_1 - \omega t)} + Ae^{i(kx_2 - \omega t)}$$

The differences from Prob. 1.2 are as follows:

(1) Here we add the complex amplitudes, not just the real part;
(2) The probability density is just the absolute square of this quantity;
(3) It is unnecessary to take a time average.

The probability at the observing screen is therefore

$$
\begin{aligned}
|\Psi(x,t)|^2 &= |A|^2 \left| e^{i(kx_1 - \omega t)} + e^{i(kx_2 - \omega t)} \right|^2 \\
&= 2|A|^2 \left\{ 1 + \cos\left[k(x_2 - x_1)\right] \right\} \\
&= 4|A|^2 \cos^2\left(\frac{k\Delta}{2}\right)
\end{aligned}
$$

The probability density for finding a particle at a given position on the observing screen looks exactly like the intensity distribution in classical optics in Prob. 1.2(a).

Problem 2.2 (a) Consider a particle of mass m_0 constrained to move in a circle of radius a in the (x, y)-plane. Show the classical hamiltonian is

$$H = \frac{1}{2I} L_z^2 \qquad ; I = m_0 a^2$$

where L_z is the angular momentum in the z-direction, and I is the moment of inertia;

(b) In quantum mechanics, L_z is given by

$$L_z = x p_y - y p_x = \frac{\hbar}{i} \left(x \frac{\partial}{\partial y} - y \frac{\partial}{\partial x} \right)$$

It is convenient to measure angular momentum in units of $\hbar$, and to define $L_z \equiv \hbar l_z$. It follows that

$$l_z = \frac{1}{i} \left(x \frac{\partial}{\partial y} - y \frac{\partial}{\partial x} \right) \qquad ; L_z \equiv \hbar l_z$$

Consider an eigenfunction of l_z with eigenvalue m

$$l_z f(x, y) = m f(x, y) \qquad ; \text{eigenvalue } m$$

Make this an implicit function of the rotation angle ϕ by defining

$$\psi(\phi) \equiv f(x(\phi), y(\phi)) \qquad ; x = a \cos \phi$$
$$; y = a \sin \phi$$

Show

$$\frac{1}{i} \frac{d\psi(\phi)}{d\phi} = \frac{1}{i} \left(x \frac{\partial}{\partial y} - y \frac{\partial}{\partial x} \right) f(x, y) = l_z f(x, y)$$

Hence conclude that the eigenvalue equation for the z-component of the angular momentum can be *rewritten* as

$$l_z \psi(\phi) \equiv \frac{1}{i} \frac{d\psi(\phi)}{d\phi} = m \psi(\phi)$$

(c) Show the eigenfunctions of l_z obeying p.b.c. are

$$\psi_m(\phi) = \frac{1}{\sqrt{2\pi}} e^{im\phi} \qquad ; m = 0, \pm 1, \pm 2, \cdots$$

(d) Show these are simultaneous eigenstates of the hamiltonian, with energy eigenvalues

$$E_m = \frac{\hbar^2}{2I} m^2$$

Solution to Problem 2.2 (a) Suppose one has a particle of mass m_0 constrained to move in a circle of radius a in the (x, y)-plane. The kinetic energy, which here is the hamiltonian, is

$$T = H = \tfrac{1}{2} m_0 v^2 = \tfrac{1}{2} m_0 (a\omega)^2$$

where ω is the angular frequency. The corresponding z-component of the angular momentum is

$$L_z = m_0 a v = m_0 a^2 \omega$$

It follows that

$$H = \frac{1}{2I} L_z^2 \qquad ; I = m_0 a^2$$

where I is the moment of inertia.

Another way to picture this is as a rigid body with the mass m_0 held at a distance a by a massless rigid rod, and free to rotate in a plane about the origin.

(b) In quantum mechanics, L_z is given by

$$L_z = x p_y - y p_x = \frac{\hbar}{i} \left(x \frac{\partial}{\partial y} - y \frac{\partial}{\partial x} \right)$$

It is convenient to measure angular momentum in units of $\hbar$, and to define $L_z \equiv \hbar l_z$. It follows that

$$l_z = \frac{1}{i} \left(x \frac{\partial}{\partial y} - y \frac{\partial}{\partial x} \right) \qquad ; L_z \equiv \hbar l_z$$

Consider an eigenfunction of l_z with eigenvalue m

$$l_z f(x, y) = m f(x, y) \qquad ; \text{eigenvalue } m$$

Make this an implicit function of the rotation angle ϕ by defining

$$\psi(\phi) \equiv f[x(\phi), y(\phi)] \qquad ; x = a \cos \phi$$
$$; y = a \sin \phi$$

Now use the rule for differentiating an implicit function to obtain

$$\frac{1}{i}\frac{d\psi(\phi)}{d\phi} = \frac{1}{i}\left(x\frac{\partial}{\partial y} - y\frac{\partial}{\partial x}\right)f(x,y) = l_z f(x,y)$$

Hence we conclude that the eigenvalue equation for the z-component of the angular momentum can be *rewritten* as

$$l_z\psi(\phi) \equiv \frac{1}{i}\frac{d\psi(\phi)}{d\phi} = m\psi(\phi)$$

(c) The solution to this equation is $\psi(\phi) = e^{im\phi}$, and if we impose p.b.c. on the interval $[0, 2\pi]$, then m must be an integer. Hence the normalized eigenfunctions of l_z obeying p.b.c. are

$$\psi_m(\phi) = \frac{1}{\sqrt{2\pi}}e^{im\phi} \qquad ; \; m = 0, \pm 1, \pm 2, \cdots$$

(d) These are evidently simultaneous eigenstates of the hamiltonian, with energy eigenvalues

$$E_m = \frac{\hbar^2}{2I}m^2$$

Problem 2.3 The results in the previous problem can be used to write the Schrödinger equation in polar coordinates in two dimensions. Introduce the coordinates (r, ϕ), and let a partial derivative in these coordinates indicate that the other variable in this pair is to be held fixed.

(a) Show from the previous problem that

$$\frac{\partial \psi(r, \phi)}{\partial \phi} = \left(x\frac{\partial}{\partial y} - y\frac{\partial}{\partial x}\right)f(x,y) \qquad ; \; x = r\cos\phi$$

$$; \; y = r\sin\phi$$

(b) In exactly the same manner, show

$$\left(r\frac{\partial}{\partial r}\right)\psi(r, \phi) = \left(x\frac{\partial}{\partial x} + y\frac{\partial}{\partial y}\right)f(x,y)$$

(c) Apply these relations twice, add, and divide by r^2 to obtain the laplacian in polar coordinates

$$\left[\frac{1}{r}\frac{\partial}{\partial r}\left(r\frac{\partial}{\partial r}\right) + \frac{1}{r^2}\frac{\partial^2}{\partial \phi^2}\right]\psi(r, \phi) = \left(\frac{\partial^2}{\partial x^2} + \frac{\partial^2}{\partial y^2}\right)f(x,y)$$

Solution to Problem 2.3 (a) It was shown in Prob. 2.2, using the derivative of an implicit function, that at fixed r

$$\frac{\partial \psi(r, \phi)}{\partial \phi} = \left(x\frac{\partial}{\partial y} - y\frac{\partial}{\partial x} \right) f(x, y) \qquad ; x = r\cos\phi$$

$$; y = r\sin\phi$$

(b) It follows in exactly the same fashion that at fixed ϕ

$$\left(r\frac{\partial}{\partial r} \right) \psi(r, \phi) = \left(x\frac{\partial}{\partial x} + y\frac{\partial}{\partial y} \right) f(x, y)$$

(c) Apply these relations twice

$$\frac{\partial^2 \psi(r, \phi)}{\partial \phi^2} = \left(x\frac{\partial}{\partial y} - y\frac{\partial}{\partial x} \right)\left(x\frac{\partial}{\partial y} - y\frac{\partial}{\partial x} \right) f(x, y)$$

$$\left(r\frac{\partial}{\partial r} \right)\left(r\frac{\partial}{\partial r} \right) \psi(r, \phi) = \left(x\frac{\partial}{\partial x} + y\frac{\partial}{\partial y} \right)\left(x\frac{\partial}{\partial x} + y\frac{\partial}{\partial y} \right) f(x, y)$$

Now add them together and divide by $r^2 = x^2 + y^2$

$$\left[\frac{1}{r}\frac{\partial}{\partial r}\left(r\frac{\partial}{\partial r} \right) + \frac{1}{r^2}\frac{\partial^2}{\partial \phi^2} \right] \psi(r, \phi) = \left(\frac{\partial^2}{\partial x^2} + \frac{\partial^2}{\partial y^2} \right) f(x, y)$$

This is the laplacian in polar coordinates.

Problem 2.4 The eigenfunctions and eigenvalues for a particle of mass m_0 constrained to move in a circle of radius a in the (x, y)-plane are given in Prob. 2.2 as

$$\psi_m(\phi) = \frac{1}{\sqrt{2\pi}}e^{im\phi} \qquad ; m = 0, \pm 1, \pm 2, \cdots$$

$$E_m = \frac{\hbar^2}{2I}m^2$$

(a) Show the corresponding probability density is a uniform constant;

(b) Show the general solution is

$$\Psi(\phi, t) = \sum_m c_m \psi_m(\phi) e^{-iE_m t/\hbar}$$

(c) Suppose one creates an initial state

$$\Psi(\phi, 0) = \frac{1}{\sqrt{2}}[\psi_{m_1}(\phi) + \psi_{m_2}(\phi)]$$

Construct the corresponding general solution, and show the probability density now oscillates as a function of time.

(d) What is the frequency of this oscillation?

Solution to Problem 2.4 As stated, the eigenfunctions and eigenvalues for a particle of mass m_0 constrained to move in a circle of radius a in the (x, y)-plane are given in Prob. 2.2 as

$$\psi_m(\phi) = \frac{1}{\sqrt{2\pi}} e^{im\phi} \qquad ; \; m = 0, \pm 1, \pm 2, \cdots$$

$$E_m = \frac{\hbar^2}{2I} m^2$$

(a) The probability density in an eigenstate is

$$|\psi_m(\phi)|^2 = \frac{1}{2\pi}$$

This is a constant independent of angle in the stationary state. There is no preferred position.

(b) The general solution to the linear Schrödinger equation is obtained by a superposition of the eigenstates with constant coefficients c_m obtained from the initial conditions

$$\Psi(\phi, t) = \sum_m c_m \psi_m(\phi) e^{-iE_m t/\hbar}$$

(c) Given the initial condition

$$\Psi(\phi, 0) = \frac{1}{\sqrt{2}} \left[\psi_{m_1}(\phi) + \psi_{m_2}(\phi) \right]$$

The general solution is immediately determined to be

$$\Psi(\phi, t) = \frac{1}{\sqrt{2}} \left[\psi_{m_1}(\phi) e^{-iE_{m_1} t/\hbar} + \psi_{m_2}(\phi) e^{-iE_{m_2} t/\hbar} \right]$$

The corresponding probability density is

$$\left| \frac{1}{\sqrt{2}} \left[\psi_{m_1}(\phi) e^{-iE_{m_1} t/\hbar} + \psi_{m_2}(\phi) e^{-iE_{m_2} t/\hbar} \right] \right|^2 =$$

$$\frac{1}{2\pi} \left[1 + \mathrm{Re} \left\{ e^{i[(m_1 - m_2)\phi - (E_{m_1} - E_{m_2})t/\hbar]} \right\} \right]$$

(d) Now use

$$\mathrm{Re} \left\{ e^{i[(m_1 - m_2)\phi - (E_{m_1} - E_{m_2})t/\hbar]} \right\} = \cos \left[(m_1 - m_2)\phi - \frac{1}{\hbar}(E_{m_1} - E_{m_2})t \right]$$

This expression oscillates with time with an angular frequency given by

$$\hbar\omega = E_{m_1} - E_{m_2}$$

At a fixed time, it also oscillates with the angle according to $(m_1 - m_2)\phi$.

The dependence on time and angle comes from the *interference* of the stationary states.

Problem 2.5 (a) Show the corresponding probability current in Prob. 2.4 is[1]

$$S(\phi, t) = \frac{1}{2m_0 a} \left\{ \Psi^\star(\phi, t) \frac{\hbar}{i} \frac{\partial \Psi(\phi, t)}{\partial \phi} + \left[\frac{\hbar}{i} \frac{\partial \Psi(\phi, t)}{\partial \phi} \right]^\star \Psi(\phi, t) \right\}$$

(b) Interpret this quantity;

(c) Calculate $S(\phi, t)$ for the solution $\Psi(\phi, t) = \psi_m(\phi) e^{-iE_m t/\hbar}$.

Solution to Problem 2.5 (a) The general expression for the probability flux in Eq. (4.28) is[2]

$$\vec{S}(\vec{x}) = \frac{\hbar}{2im} \left[\psi^\star \vec{\nabla} \psi - \left(\vec{\nabla} \psi \right)^\star \psi \right]$$

For the particle moving around in a circle, we compute the component of the gradient pointing in the direction of the tangent, with a magnitude of the displacement of $ds = a\,d\phi$. Thus the probability current in the direction of increasing ϕ is

$$S(\phi, t) = \frac{1}{2m_0 a} \left\{ \Psi^\star(\phi, t) \frac{\hbar}{i} \frac{\partial \Psi(\phi, t)}{\partial \phi} + \left[\frac{\hbar}{i} \frac{\partial \Psi(\phi, t)}{\partial \phi} \right]^\star \Psi(\phi, t) \right\}$$

(b) This is the amount of probability flowing past a given point on the circle per unit time.

(c) We can immediately calculate $S(\phi, t)$ for the solution

$$\Psi(\phi, t) = \psi_m(\phi) e^{-iE_m t/\hbar} \qquad ; \ \psi_m(\phi) = \frac{1}{\sqrt{2\pi}} e^{im\phi}$$

The result is

$$S(\phi, t) = \frac{\hbar m}{2\pi m_0 a}$$

---

[1]Recall Eq. (4.28); here m_0 is the rest mass, and ϕ is the polar angle.
[2]See Prob. 4.5.

Although the probability density is constant around the circle and we do not know where the particle is, the probability current flows around the circle, tracking the motion of the particle.

If we write the angular momentum as $L_z = m_0 v a = \hbar m$, then

$$S(\phi, t) = \frac{v}{2\pi} \qquad ; \; L_z = m_0 v a$$
$$= v |\psi_m(\phi)|^2$$

The probability flux is the local probability density (a constant here) times the particle velocity.

Problem 2.6 The mean value of a hermitian operator O with a normalized wave function $\psi(x)$ is

$$\langle O \rangle = \int dx \, \psi^*(x) O \psi(x)$$

In quantum mechanics we are dealing with probability distributions, and the *mean-square-deviation* from this mean value is given by

$$(\Delta O)^2 \equiv \left\langle (O - \langle O \rangle)^2 \right\rangle = \langle O^2 \rangle - \langle O \rangle^2$$

Make use of the normalized ground-state wave function for a particle in a one-dimensional box of length L in Eq. (3.12) (recall Figs. 3.2 and 3.3 in the text), and demonstrate the following:

(a) Show that for the momentum

$$\langle p \rangle = 0 \qquad ; \; \langle p^2 \rangle = \hbar^2 \left(\frac{\pi}{L} \right)^2$$

(b) Show that for the spatial coordinate[3]

$$\langle x \rangle = \frac{L}{2} \qquad ; \; \langle x^2 \rangle = \frac{L^2}{3} \left(1 - \frac{3}{2\pi^2} \right)$$

(c) Combine these results to obtain

$$(\Delta p)^2 (\Delta x)^2 = \frac{\hbar^2 \pi^2}{12} \left(1 - \frac{6}{\pi^2} \right)$$
$$\Delta p \, \Delta x = 0.568 \, \hbar$$

---

[3]Note the following definite integrals

$$\int_0^\pi du \, u \sin^2(u) = \frac{\pi^2}{4} \qquad ; \int_0^\pi du \, u^2 \sin^2(u) = \frac{\pi^3}{6} - \frac{\pi}{4}$$

This is an example of the Heisenberg uncertainty principle. The momentum and position (p, q) of a particle *cannot both be specified precisely*. Given the commutation relation $[p, q] = \hbar/i$, it is possible to give a rigorous proof that for a normalized wave function[4]

$$\Delta p\, \Delta q \geq \frac{1}{2}\hbar \qquad ; \text{ uncertainty principle}$$

The uncertainty principle represents a significant break from classical mechanics where one initializes a mechanical system at a given point (p, q) in phase space, and then follows it in a deterministic fashion from there.

Solution to Problem 2.6 We calculate the square of the expectation value of the operator, and the expectation value of its square, using the wave function for the ground state of a particle in a one-dimensional box of length L in Eq. (3.12)

$$\psi_1(x) = \sqrt{\frac{2}{L}} \sin\left(\frac{\pi x}{L}\right)$$

(a) Then, for the coordinate $p = (\hbar/i)\partial/\partial x$

$$\langle p \rangle = \frac{\hbar}{i}\left(\frac{2}{L}\right) \int_0^L dx\, \sin\left(\frac{\pi x}{L}\right) \frac{d}{dx} \sin\left(\frac{\pi x}{L}\right) = 0$$

$$\langle p^2 \rangle = \hbar^2 \left(\frac{\pi}{L}\right)^2 \int_0^L dx\, \psi_1^*(x)\psi_1(x) = \hbar^2 \left(\frac{\pi}{L}\right)^2$$

(b) For the coordinate x, with the aid of the two definite integrals in the footnote, we have

$$\langle x \rangle = \frac{2}{L} \int_0^L dx\, x \sin^2\left(\frac{\pi x}{L}\right) = \frac{2}{L}\left(\frac{L}{\pi}\right)^2 \int_0^\pi du\, u \sin^2(u)$$
$$= \left(\frac{2L}{\pi^2}\right)\frac{\pi^2}{4} = \frac{L}{2}$$

and

$$\langle x^2 \rangle = \frac{2}{L} \int_0^L dx\, x^2 \sin^2\left(\frac{\pi x}{L}\right) = \frac{2}{L}\left(\frac{L}{\pi}\right)^3 \int_0^\pi du\, u^2 \sin^2(u)$$
$$= \left(\frac{2L^2}{\pi^3}\right)\left(\frac{\pi^3}{6} - \frac{\pi}{4}\right) = \frac{L^2}{3}\left(1 - \frac{3}{2\pi^2}\right)$$

---

[4] See, for example, [Walecka (2008)].

(c) These results can be combined to give

$$(\Delta p)^2 = \langle p^2 \rangle - \langle p \rangle^2 = \hbar^2 \left(\frac{\pi}{L} \right)^2$$

$$(\Delta x)^2 = \langle x^2 \rangle - \langle x \rangle^2 = \frac{L^2}{3} \left(1 - \frac{3}{2\pi^2} \right) - \left(\frac{L}{2} \right)^2$$

$$= \frac{L^2}{12} \left(1 - \frac{6}{\pi^2} \right)$$

Hence

$$(\Delta p)^2 (\Delta x)^2 = \frac{\hbar^2 \pi^2}{12} \left(1 - \frac{6}{\pi^2} \right)$$

$$\Delta p \, \Delta x = 0.568 \, \hbar$$

As stated in the problem, this is an example of the *Heisenberg uncertainty principle*. The momentum and position (p, q) of a particle *cannot both be specified precisely*. If you want to localize the particle, a spread in wavenumbers, or momenta, is required in the wave function (see the next problem), and if you want to specify the momentum of the particle precisely, you do not know where it is (see Sec. 2.5).

Given the commutation relation $[p, q] = \hbar/i$, it is possible to give a rigorous proof that for a normalized wave function[5]

$$\Delta p \, \Delta q \geq \frac{1}{2} \hbar \qquad ; \text{ uncertainty principle}$$

Indeed, the result for the ground state in the box just slightly exceeds the value $\hbar/2$.

Again, as stated, the uncertainty principle represents a significant break from classical mechanics where one initializes a mechanical system at a given point (p, q) in phase space, and then follows it in a deterministic fashion from there.

Problem 2.7 Although the solutions to the Schrödinger equation and corresponding probability densities for a particle in a box in Figs. 3.2 and 3.3 in the text are so simple and clear, the solution for a free particle is more subtle.[6] For example, the probability density obtained from our motivating wave in Eq. (2.5) is *independent of position and time!* In order to construct a probability density that moves with the particle, we need to construct a *wave packet*.

---

[5]See, for example, [Walecka (2008)].
[6]This is why we have left it as a problem—so as not to interrupt the flow of the text.

(a) Take a *superposition* of the waves in Eq. (2.5) with an amplitude sharply peaked about a wave number corresponding to the classical momentum $p_0 = \hbar k_0$.[7] Construct

$$\Psi(x,t) = \int dk\, A(k-k_0)e^{i[kx - \omega(k)t]} \qquad ; \omega(k) = \hbar k^2/2m$$

Show this satisfies the Schrödinger equation for a free particle;

(b) Show that at the initial time $t = 0$, the wave function and probability density are

$$\Psi(x,0) = e^{ik_0 x} F(x) \qquad ; |\Psi(x,0)|^2 = |F(x)|^2$$

$$F(x) = \int dl\, A(l)\, e^{ilx}$$

where the amplitude $A(l) = A(k - k_0)$ is sharply peaked about $l = 0$;

(c) As an example, take $A(l) = 1$ for $|l| < l_0$; calculate and plot $|F(x)|^2$;

(d) Since $A(l)$ is sharply peaked about $l = 0$, we can make a Taylor series expansion of $\omega(k)$ about k_0 and keep just the first term

$$\omega(k) \approx \omega(k_0) + (k - k_0)\left[\frac{d\omega(k)}{dk}\right]_{k_0}$$

Show that for finite time, the probability density then becomes

$$|\Psi(x,t)|^2 = |F(x - v_{\mathrm{gp}}t)|^2$$

Here v_{gp} is the *group velocity*[8]

$$v_{\mathrm{gp}} = \left[\frac{d\omega(k)}{dk}\right]_{k_0} = \frac{\hbar k_0}{m} = \frac{p_0}{m}$$

Thus the probability density for the localized wave packet moves with the classical particle velocity!

Solution to Problem 2.7 (a) In order to construct a localized wave packet, we take a linear *superposition* of the dispersive motivating wave in Eq. (2.5)

$$\Psi(x,t) = \int dk\, A(k-k_0)e^{i[kx-\omega(k)t]} \qquad ; \omega(k) = \frac{\hbar k^2}{2m}$$

---

[7]Recall the previous problem on the uncertainty principle.
[8]The *phase velocity* of a wave is $v_{\mathrm{ph}} = \omega(k)/k$. Note that for this free-particle wave packet, the group velocity is twice the phase velocity.

If the integral is convergent, we are free to differentiate under the integral sign, and this expression obeys the Schrödinger equation exactly as demonstrated in the text.

(b) Consider the initial time $t = 0$. With the definition $l \equiv k - k_0$, the wave function and probability density follow as

$$\Psi(x, 0) = e^{ik_0 x} F(x)$$

$$F(x) = \int dl \, A(l) \, e^{ilx}$$

where the amplitude $A(l) = A(k - k_0)$ is sharply peaked about $l = 0$. The probability density follows as

$$|\Psi(x, 0)|^2 = |F(x)|^2$$

(c) As an example, we take $A(l) = 1$ for $|l| < l_0$. The expression for $F(x)$ then follows as

$$F(x) = \int_{-l_0}^{l_0} dl \, e^{ilx} = \frac{1}{ix} \left(e^{il_0 x} - e^{-il_0 x} \right) = \frac{2}{x} \sin(l_0 x)$$

The square of the modulus of this expression is

$$|F(x)|^2 = 4l_0^2 \left[\frac{\sin(l_0 x)}{l_0 x} \right]^2$$

If we compare this with Eq. (5.21), we can write

$$\frac{1}{K} |F(x)|^2 = f_K(x) \qquad ; \ K \equiv 2l_0$$

With the appropriate transcription $(T \to K, \, \omega \to x)$, the quantity $f_K(x)$ is plotted in Fig. 5.1 in the text. The packet is well localized, with the first zero of $f_K(x)$ coming at $x = 2\pi/K$. The bigger K, the more localized is the packet.

(d) Since $A(l)$ is sharply peaked about $l = 0$, we can make a Taylor series expansion of $\omega(k)$ about k_0 and keep just the first term

$$\omega(k) \approx \omega(k_0) + (k - k_0) \left[\frac{d\omega(k)}{dk} \right]_{k_0}$$

Just as in part (b), the wave function can then be written as

$$\Psi(x,t) \approx \int dk \, A(k-k_0) e^{i[kx-\omega(k_0)t-(k-k_0)v_{gp}t]}$$

$$= e^{i[k_0 x - \omega(k_0)t]} \int dl \, A(l) e^{il(x-v_{gp}t)}$$

$$= e^{i[k_0 x - \omega(k_0)t]} F(x - v_{gp}t)$$

where v_{gp} is the *group velocity* of the wave

$$v_{gp} = \left[\frac{d\omega(k)}{dk}\right]_{k_0} = \frac{\hbar k_0}{m} = \frac{p_0}{m}$$

This is the classical velocity of a particle with momentum p_0.

The corresponding probability density for the wave packet is[9]

$$|\Psi(x,t)|^2 = |F(x - v_{gp}t)|^2$$

From our discussion of wave motion, this expression represents a wave packet moving with the group velocity v_{gp} in the positive x-direction, with no change in shape. Thus the probability density for this localized wave packet just moves with the classical particle velocity! [10]

---

[9] Notice how the carrier wave factors out and cancels in the probability density.

[10] If the full expression is kept for $\omega(k)$, there will be some spreading of the wave packet (see [Walecka (2008)]).

Chapter 3

Include Potential $V(x)$

Problem 3.1 Suppose the potential in Fig. 3.4 in the text has the opposite sign and is an *attractive half-space*, with $V(x) = -V_0 < 0$ for all positive x. Write the transmitted wave as

$$\psi(x) = t\,e^{i\kappa x} \qquad ; \ \kappa^2 = \frac{2m}{\hbar^2}(E + V_0)$$

Match the boundary conditions at $x = 0$, and show the transmitted amplitude is

$$t = \frac{2k}{k + \kappa}$$

Discuss.

Solution to Problem 3.1 The Schrödinger equation in the attractive half-space reads

$$\left(\frac{d^2}{dx^2} + \kappa^2\right)\psi(x) = 0 \qquad ; \ \kappa^2 = \frac{2m}{\hbar^2}(E + V_0)$$

We are then looking for a solution of the form

$$\psi(x) = e^{ikx} + re^{-ikx} \qquad ; \ x < 0$$
$$= te^{i\kappa x} \qquad ; \ x > 0$$

Now match the wave functions and the derivatives at the origin

$$1 + r = t$$
$$ik(1 - r) = i\kappa t$$

The solution to these equations is

$$t = \frac{2k}{k+\kappa} \qquad ; \, r = \frac{k-\kappa}{k+\kappa}$$

We note the following:

- For very large κ, or a very deep potential $-V_0$, the transmitted amplitude t goes to zero. There is no transmission into the deep well! *This is a quite amazing result of quantum mechanics.*
- For large κ, the reflected amplitude $r = -1$, and the wave function on the left vanishes at the potential. The very deep attractive potential acts as a wall!
- In this limit, the reflected wave has a phase change of π from the incident wave $(r = -1)$.

Problem 3.2 Show that the expansion coefficients in the general solution for a particle in a square box in Eq. (3.42) are obtained from the initial condition

$$\Psi(x, y, 0) = g(x, y)$$

according to

$$c_{n_x, n_y} = \int_0^L dx \int_0^L dy \, \psi^*_{n_x, n_y}(x, y) \, g(x, y)$$

Solution to Problem 3.2 The eigenfunctions and eigenvalues for the particle in a two-dimensional square box are given in Eqs. (3.41)

$$\psi_{n_x, n_y}(x, y) = \left(\frac{2}{L}\right) \sin\left(\frac{n_x \pi x}{L}\right) \sin\left(\frac{n_y \pi y}{L}\right) \qquad ; \, (n_x, n_y) = 1, 2, 3, \cdots$$

$$E_{n_x, n_y} = \frac{\hbar^2 \pi^2}{2mL^2} \left(n_x^2 + n_y^2\right)$$

The general solution to the Schrödinger equation is correspondingly

$$\Psi(x, y, t) = \sum_{n_x} \sum_{n_y} c_{n_x, n_y} \psi_{n_x, n_y}(x, y) \, e^{-iE_{n_x, n_y} t/\hbar}$$

The expansion coefficients c_{n_x, n_y} can be obtained from the initial condition

$$\Psi(x, y, 0) = g(x, y)$$

Write

$$g(x, y) = \sum_{n_x} \sum_{n_y} c_{n_x, n_y} \psi_{n_x, n_y}(x, y)$$

$$= \sum_{n_x} \sum_{n_y} c_{n_x, n_y} \frac{2}{L} \sin\left(\frac{n_x \pi x}{L}\right) \sin\left(\frac{n_y \pi y}{L}\right)$$

Now use the orthonormality of each of the one-dimensional solutions to arrive at

$$c_{n_x, n_y} = \frac{2}{L} \int_0^L dx \int_0^L dy \sin\left(\frac{n_x \pi x}{L}\right) \sin\left(\frac{n_y \pi y}{L}\right) g(x, y)$$

$$= \int_0^L dx \int_0^L dy \, \psi^*_{n_x, n_y}(x, y) \, g(x, y)$$

Problem 3.3 (a) Suppose one prepares the following initial state for the particle in the one-dimensional box[1]

$$\Psi(x, 0) = \frac{1}{\sqrt{2}} \left[\psi_1(x) + \psi_2(x)\right]$$

Plot the initial wave function and probability distribution.

(b) Construct the solution $\Psi(x, t)$ and probability distribution $|\Psi(x, t)|^2$ for later times;

(c) Show that probability distribution oscillates back and forth in the box;

(d) What is the frequency of that oscillation?

Solution to Problem 3.3 The wave functions and energies for the particle in a one-dimensional box are given in Eqs. (3.12) and (3.13)

$$\psi_n(x) = \sqrt{\frac{2}{L}} \sin\left(\frac{n \pi x}{L}\right) \qquad ; n = 1, 2, 3, \cdots$$

$$E_n = \frac{\hbar^2}{2m} \left(\frac{n\pi}{L}\right)^2$$

The general solution to the Schrödinger equation is

$$\Psi(x, t) = \sum_n c_n \psi_n(x) e^{-iE_n t/\hbar}$$

---

[1] See Figs. 3.2 and 3.3 in the text.

where the c_n are constants determined from the initial condition

$$\Psi(x,0) = \sum_n c_n \psi_n(x)$$

(a) With the given initial condition, the only non-zero coefficients are

$$c_1 = c_2 = \frac{1}{\sqrt{2}}$$

The initial wave function is therefore

$$\Psi(x,0) = \frac{1}{\sqrt{L}} \left[\sin\left(\frac{\pi x}{L}\right) + \sin\left(\frac{2\pi x}{L}\right) \right]$$

and the initial probability density is $|\Psi(x,0)|^2$, which is readily plotted (see below)

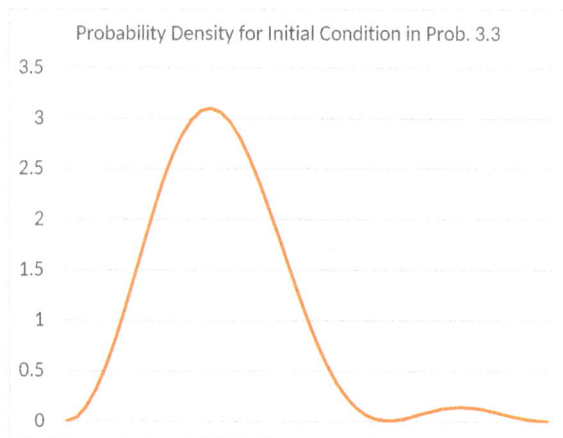

Probability Density for Initial Condition in Prob. 3.3

Fig. 3.1   Initial probability density $L|\Psi(x,0)|^2$ as a function of x/L between $(0,1)$ for the particle in a box with the initial conditions of Prob. 3.3.

(b) The general solution for $t > 0$ in this case is

$$\Psi(x,t) = \frac{1}{\sqrt{L}} \left[\sin\left(\frac{\pi x}{L}\right) e^{-iE_1 t/\hbar} + \sin\left(\frac{2\pi x}{L}\right) e^{-iE_2 t/\hbar} \right]$$

(c) The corresponding probability distribution is

$$|\Psi(x,t)|^2 = \frac{1}{L}\left\{ \sin^2\left(\frac{\pi x}{L}\right) + \sin^2\left(\frac{2\pi x}{L}\right) + \right.$$
$$\left. 2\sin\left(\frac{\pi x}{L}\right)\sin\left(\frac{2\pi x}{L}\right)\cos\left[\frac{1}{\hbar}(E_2 - E_1)t\right] \right\}$$

This probability density oscillates in time as the particle bounces back and forth in the box.

(d) The angular frequency of the oscillation of the probability density is[2]

$$\hbar\omega = E_2 - E_1 = \frac{\hbar^2}{2m}\left[\left(\frac{2\pi}{L}\right)^2 - \left(\frac{\pi}{L}\right)^2\right]$$

Problem 3.4 Suppose one prepares an initial state for the particle in a box that is simply constant over the box

$$\Psi(x,0) = \frac{1}{\sqrt{L}} \qquad ; \ 0 \leq x \leq L$$

Show the solution to the Schrödinger equation for all subsequent time is

$$\Psi(x,t) = \sum_{n=1}^{\infty} c_n\psi_n(x)e^{-iE_n t/\hbar}$$
$$c_n = \frac{\sqrt{2}}{\pi}\left[\frac{1 - (-1)^n}{n}\right]$$

It is interesting that this simplest of initial conditions gives rise to such a complicated wave function.

Solution to Problem 3.4 The general solution for a particle in a box in one dimension is given by linear superposition as

$$\Psi(x,t) = \sum_{n=1}^{\infty} c_n\psi_n(x)e^{-iE_n t/\hbar}$$

where the $\psi_n(x)$ are the orthonormal eigenfunctions in Eqs. (3.12). The coefficients c_n are then obtained from the initial condition as

$$c_n = \int_0^L dx\,\psi_n^*(x)\Psi(x,0)$$

---

[2] After a time $t = \pi/\omega$, the probability density peaks in the r.h.s. of the box (plot it!).

With the given initial condition, this yields

$$c_n = \frac{1}{\sqrt{L}}\sqrt{\frac{2}{L}}\int_0^L dx \sin\left(\frac{n\pi x}{L}\right) \qquad ; \ n = 1, 2, 3, \cdots$$

The integral is immediately evaluated as

$$c_n = \frac{\sqrt{2}}{L}\frac{1}{n\pi/L}\left[-\cos\left(\frac{n\pi x}{L}\right)\right]_0^L$$

$$= \frac{\sqrt{2}}{n\pi}[1 - \cos(n\pi)]$$

Hence

$$c_n = \frac{\sqrt{2}}{\pi}\frac{[1-(-1)^n]}{n} \qquad ; \ n = 1, 2, 3, \cdots$$

Problem 3.5 Suppose there is a small circular potential at the center of the two-dimensional square box of the form

$$\delta V(\vec{r}) = v_0 \qquad ; \ |\vec{r} - \vec{r}_0| < a$$

where $\vec{r}_0$ is located at the center of the box. Assume $a \ll L$. Use perturbation theory to show that the shift in the ground-state eigenvalue is then

$$\delta E_{1,1} = 4v_0 \frac{\pi a^2}{L^2}$$

Solution to Problem 3.5 First-order perturbation theory gives the shift in the ground-state energy eigenvalue as

$$\delta E_{1,1} = \int_0^L dx \int_0^L dy \ \psi_{1,1}^\star(x,y)\delta V(\vec{r})\psi_{1,1}(x,y)$$

where $\psi_{1,1}(x,y)$ is the normalized ground-state wave function

$$\psi_{1,1}(x,y) = \frac{2}{L}\sin\left(\frac{\pi x}{L}\right)\sin\left(\frac{\pi y}{L}\right)$$

At the center of the box, the absolute square of the wave function is simply

$$\left|\psi_{1,1}\left(\frac{L}{2},\frac{L}{2}\right)\right|^2 = \frac{4}{L^2}$$

The perturbation is

$$\delta V(\vec{r}) = \nu_0 \qquad ; |\vec{r} - \vec{r}_0| < a$$

where $\vec{r}_0$ is located at the center of the box. If a is small enough so that the square of the wave function does not vary much over the perturbation, then the above value of that square of the wave function can be factored from the integral, and it simply becomes the *area* of the perturbation

$$\delta E_{1,1} \approx \nu_0 \frac{4}{L^2} \pi a^2$$

It follows that for $a \ll L$

$$\delta E_{1,1} = 4\nu_0 \frac{\pi a^2}{L^2} \qquad ; a \ll L$$

Problem 3.6 Consider the non-degenerate perturbation theory in Eqs. (3.55).

(a) Show that this analysis holds for a particle in a one-dimensional box with an additional potential $\delta V(x)$;

(b) Suppose the perturbation $\delta V(x)$ is *odd* about the midpoint of the box. Show that all the first-order energy shifts then vanish;

(c) Show that the second-order energy shift always *lowers* the energy of the ground state.

Solution to Problem 3.6 (a) The energy levels for a particle in a one-dimensional box are all distinct; every state has a different energy

$$\psi_n(x) = \sqrt{\frac{2}{L}} \sin\left(\frac{n\pi x}{L}\right) \qquad ; n = 1, 2, 3, \cdots$$

$$E_n = \frac{\hbar^2}{2m}\left(\frac{n\pi}{L}\right)^2$$

Every energy is different in $\sum_{m \neq n}$ in Eqs. (3.55), and therefore, for a perturbation in the one-dimensional box, one can then do non-degenerate perturbation theory.

(b) The square of the wave functions $|\psi_n(x)|^2$ is even about the midpoint of the box (see Fig. 3.3 in the text). The integral of this expression multiplied by a function $\delta V(x)$ that is *odd* about the midpoint of the box then gives zero. All the first-order energy shifts thus vanish

$$\int_0^L dx\, |\psi_n(x)|^2 \delta V(x) = 0 \qquad ; \delta V\left(\frac{L}{2} - x\right) = -\delta V\left(\frac{L}{2} + x\right)$$

(c) The second-order energy shift in Eqs. (3.55) is

$$\delta E_n^{(2)} = \sum_{m \neq n} \frac{1}{E_n^0 - E_m^0} \left| \int dy\, \psi_m^*(y)\delta V(y)\psi_n(y) \right|^2$$

If $n = 1$ is the ground state, then all the energy denominators in this expression are negative, as are then all the terms in the sum. Hence the second-order energy shift always *lowers* the energy of the ground state.

Chapter 4

Scattering

Problem 4.1 Suppose the spherical square-well potential in Chapter 4 is just deep enough to have one bound state at $k^2 = 0$.

(a) Show the depth of the potential is $2mV_0/\hbar^2 = \pi^2/4d^2$;

(b) What is the s-wave wave function inside the potential?

(c) What is it outside?

Solution to Problem 4.1 Consider the spherical square-well potential studied in Secs. 4.1–4.2, and let us examine a *bound-state* with energy

$$\frac{2mE}{\hbar^2} = -\varepsilon_b$$

The s-wave Schrödinger equation now reads

$$\left[\frac{d^2}{dr^2} + v_0 - \varepsilon_b\right] u(r) = 0$$

where the s-wave wave function is

$$\psi(r) \equiv \frac{u(r)}{r}$$

Define $\kappa^2 \equiv v_0 - \varepsilon_b$. Then inside the potential with $r < d$

$$\left[\frac{d^2}{dr^2} + \kappa^2\right] u(r) = 0 \qquad ; \, \kappa^2 \equiv v_0 - \varepsilon_b$$

The solution to this equation is

$$u(r) = A \sin \kappa r + B \cos \kappa r \qquad ; \, r < d$$

Since $\psi(r) = B/r$ is too singular at the origin, one must choose $B = 0$.[1]
Hence

$$u(r) = A \sin \kappa r \qquad ; r < d$$

Outside the potential where $v_0 = 0$, one must retain just the decreasing
exponential, and hence

$$u(r) = Ce^{-\sqrt{\varepsilon_b}r} \qquad ; r > d$$

Now equate the logarithmic derivatives of these solutions at $r = d$

$$\kappa \cot \kappa d = -\sqrt{\varepsilon_b}$$

If there is just one bound state at energy $\varepsilon_b = 0$, this equation becomes

$$\cot \sqrt{v_0}\, d = 0 \qquad ; \varepsilon_b \to 0$$

In this case

$$\sqrt{v_0}\, d = \frac{\pi}{2}$$

which can be rewritten

$$\frac{2mV_0}{\hbar^2} = \frac{\pi^2}{4d^2} \qquad ; \varepsilon_b \to 0$$

The volume element in three dimensions in spherical coordinates for a
wave function that has no angle dependence is $d^3x \to 4\pi r^2 dr$. Hence the
normalization condition for the s-wave bound state is

$$\int_0^\infty |\psi(r)|^2\, 4\pi r^2 dr = 4\pi \int_0^\infty |u(r)|^2 dr = 1$$

To obtain some physical insight, we examine the unnormalized wave
function in the limit $\varepsilon_b \to 0$.

(b) The wave function *inside* the potential in this case is

$$u(r) = \sin\left(\frac{\pi r}{2d}\right) \qquad ; r < d$$

It vanishes at the origin and has zero slope at the potential boundary; there
is one-quarter wavelength inside the potential.

(c) The wave function *outside* the potential is

$$u(r) = e^{-\sqrt{\varepsilon_0}r}$$

---

[1]With the singular solution and superposition, the origin can become a point source
of probability, which is unacceptable (see [Walecka (2008)]).

This is a very slowly decreasing function of the radial coordinate; all of the wave function is outside of the potential, and in the limit $\varepsilon_b = 0$, it is just $u(r) = 1$!

Problem 4.2 In the separated radial Schrödinger equation for a free particle in spherical coordinates, there are two types of solutions that form a fundamental system in which any radial solution can be expanded. These are the spherical Bessel functions $j_l(\rho)$ and *spherical Neumann functions* $n_l(\rho)$.

(a) For $l = 0$ the spherical Neumann function is

$$n_0(\rho) = -\frac{\cos\rho}{\rho}$$

Show this satisfies the same radial equation as $j_0(\rho)$;

(b) For $l = 1$ the spherical Neumann function is

$$n_1(\rho) = -\frac{\cos\rho}{\rho^2} - \frac{\sin\rho}{\rho}$$

Show this satisfies the same radial equation as $j_1(\rho)$;

(c) Show that through this order, the Neumann functions satisfy the general relations

$$n_l(\rho) \to -\frac{1 \cdot 1 \cdot 3 \cdots (2l-1)}{\rho^{l+1}} \qquad ; \rho \to 0$$

$$n_l(\rho) \to \frac{1}{\rho} \sin\left[\rho - (l+1)\pi/2\right] \qquad ; \rho \to \infty$$

Note that the spherical Neumann functions are singular at the origin.

Solution to Problem 4.2 The spherical Bessel functions satisfy the radial equation

$$\frac{d^2 j_l(\rho)}{d\rho^2} + \frac{2}{\rho}\frac{dj_l(\rho)}{d\rho} + \left[1 - \frac{l(l+1)}{\rho^2}\right] j_l(\rho) = 0$$

(a) Consider the spherical Neumann function with $l = 0$. Then

$$n_0(\rho) = -\frac{\cos\rho}{\rho}$$

$$\frac{dn_0(\rho)}{d\rho} = \frac{\sin\rho}{\rho} + \frac{\cos\rho}{\rho^2}$$

$$\frac{d^2 n_0(\rho)}{d\rho^2} = \frac{\cos\rho}{\rho} - 2\frac{\sin\rho}{\rho^2} - 2\frac{\cos\rho}{\rho^3}$$

Hence

$$\frac{d^2 n_0(\rho)}{d\rho^2} + \frac{2}{\rho}\frac{dn_0(\rho)}{d\rho} + n_0(\rho) = 0$$

(b) Similarly, consider the spherical Neumann function for $l = 1$

$$n_1(\rho) = -\frac{\cos\rho}{\rho^2} - \frac{\sin\rho}{\rho}$$

$$\frac{dn_1(\rho)}{d\rho} = 2\frac{\sin\rho}{\rho^2} + 2\frac{\cos\rho}{\rho^3} - \frac{\cos\rho}{\rho}$$

$$\frac{d^2 n_1(\rho)}{d\rho^2} = 3\frac{\cos\rho}{\rho^2} - 6\frac{\sin\rho}{\rho^3} - 6\frac{\cos\rho}{\rho^4} + \frac{\sin\rho}{\rho}$$

Hence

$$\frac{d^2 n_1(\rho)}{d\rho^2} + \frac{2}{\rho}\frac{dn_1(\rho)}{d\rho} + \left[1 - \frac{2}{\rho^2}\right]n_1(\rho) = 0$$

(c) As $\rho \to 0$,

$$n_0(\rho) \to -\frac{1}{\rho} \qquad ; \; n_1(\rho) \to -\frac{1}{\rho^2}$$

and as $\rho \to \infty$,

$$n_0(\rho) \to -\frac{\cos\rho}{\rho} \qquad ; \; n_1(\rho) \to -\frac{\sin\rho}{\rho}$$

Problem 4.3 The spherical harmonics are the non-singular solutions to the angular part of the Schrödinger equation in spherical coordinates. For $l = 1$, one has

$$Y_{1,0}(\theta, \phi) = \sqrt{\frac{3}{4\pi}}\cos\theta \qquad ; \; Y_{1,\pm1}(\theta, \phi) = \mp\sqrt{\frac{3}{8\pi}}\sin\theta\, e^{\pm i\phi}$$

Show these satisfy the angular equation

$$\left[\frac{1}{\sin\theta}\frac{\partial}{\partial\theta}\left(\sin\theta\frac{\partial}{\partial\theta}\right) + \frac{1}{\sin^2\theta}\frac{\partial^2}{\partial\phi^2}\right]Y_{l,m}(\theta, \phi) = -l(l+1)Y_{l,m}(\theta, \phi)$$

Solution to Problem 4.3 Let us do the first one. We can cancel the overall norm and just consider

$$\frac{1}{\sin\theta}\frac{\partial}{\partial\theta}\left(\sin\theta\frac{\partial}{\partial\theta}\right)\cos\theta = -\frac{1}{\sin\theta}\frac{\partial}{\partial\theta}\sin^2\theta = -2\cos\theta$$

which is the answer.

For the second, after again cancelling the norm and phi-phase,

$$\left[\frac{1}{\sin\theta}\frac{\partial}{\partial\theta}\left(\sin\theta\frac{\partial}{\partial\theta}\right) - \frac{1}{\sin^2\theta}\right]\sin\theta = \frac{1}{\sin\theta}\left[\frac{1}{2}\frac{\partial}{\partial\theta}\sin 2\theta - 1\right]$$

$$= \frac{1}{\sin\theta}[\cos 2\theta - 1] = \frac{1}{\sin\theta}[-2\sin^2\theta] = -2\sin\theta$$

which is again the correct answer.

We make the following observation after these two problems. The Schrödinger equation outside of the spherical potential $v(r)$ reads $(\nabla^2 + k^2)\psi = 0$. The separated solutions in spherical coordinates, which are non-singular in the angles, are given by $\psi = [aj_l(kr) + bn_l(kr)]Y_{l,m}(\theta,\phi)$.

Problem 4.4 This problem involves the explicit verification of some results on the probability flux quoted in the text:

(a) Verify Eqs. (4.30);
(b) Verify Eq. (5.36).

Solution to Problem 4.4 The probability flux in three dimensions is given in Eq. (4.28)

$$\vec{S}(\vec{x}) = \frac{\hbar}{2im}\left[\psi^*\vec{\nabla}\psi - \left(\vec{\nabla}\psi\right)^*\psi\right]$$

(a) The asymptotic form of the wave function in the scattering problem is given in Eq. (4.21)

$$\psi(\vec{x}) = e^{i\vec{k}\cdot\vec{x}} + f(k,\theta)\frac{e^{ikr}}{r} \qquad ; r \to \infty$$

We first compute the flux from the incident wave

$$\psi_{\text{inc}}(\vec{x}) = e^{i\vec{k}\cdot\vec{x}}$$

This gives[2]

$$\hat{k}\cdot\vec{S}_{\text{inc}} = \hat{k}\cdot\frac{\hbar}{2im}\left[\left(e^{-i\vec{k}\cdot\vec{x}}\right)\left(i\vec{k}\,e^{i\vec{k}\cdot\vec{x}}\right) - \left(i\vec{k}\,e^{i\vec{k}\cdot\vec{x}}\right)^*\left(e^{i\vec{k}\cdot\vec{x}}\right)\right]$$

$$= \frac{\hbar k}{m}$$

Next, we compute the flux from the scattered wave

$$\psi_{\text{scatt}}(\vec{x}) = f(k,\theta)\frac{e^{ikr}}{r}$$

---

[2] In this problem $(\hat{k}, \hat{r})$ are unit vectors.

We here make use of the fact that for a function of the radial coordinate, $\hat{r} \cdot \vec{\nabla} F(r) = dF(r)/dr$. We also ignore derivatives of the denominator $1/r$, since they are negligible in the limit $r \to \infty$. It follows that

$$\hat{r} \cdot \vec{S}_{\text{scatt}} \doteq \frac{\hbar}{2imr^2} |f(k,\theta)|^2 \left[\left(e^{-ikr} \right) \left(ik\, e^{ikr} \right) - \left(ik\, e^{ikr} \right)^{\star} \left(e^{ikr} \right) \right]$$

$$= \frac{\hbar k}{mr^2} |f(k,\theta)|^2$$

(b) The one-dimensional incident wave in Eq. (5.35) is

$$\psi_{n_1^0}(x_1) = \frac{1}{\sqrt{L}} e^{ik_0 x_1}$$

The corresponding one-dimensional incident flux has just the x_1-component of the gradient. It is given by

$$I_{\text{inc}} = \frac{\hbar}{2im_1} \frac{1}{L} \left[\left(e^{-ik_0 x_1} \right) \left(ik_0\, e^{ik_0 x_1} \right) - \left(ik_0\, e^{ik_0 x_1} \right)^{\star} \left(e^{ik_0 x_1} \right) \right]$$

$$= \frac{1}{L} \left(\frac{\hbar k_0}{m_1} \right)$$

Problem 4.5 Suppose one has a Yukawa potential of the form

$$v(r) = \lambda \frac{e^{-\mu r}}{r} \qquad ; \text{ Yukawa potential}$$

Show that Born approximation for the scattering amplitude is given by

$$f_{BA}(k,\theta) = -\frac{\lambda}{q^2 + \mu^2}$$

Sketch and discuss.

Solution to Problem 4.5 First Born approximation for the scattering amplitude is given by Eq. (4.43) as[3]

$$f_{BA}(k,\theta) = -\frac{1}{4\pi} \int e^{i\vec{q}\cdot\vec{r}} v(r) d^3r \qquad ; \text{ Born approximation}$$

Let us do the integral over the angles

$$\int e^{i\vec{q}\cdot\vec{r}} d\Omega = \int_0^{2\pi} d\phi \int_{-1}^1 d\cos\theta\, e^{iqr\cos\theta}$$

$$= \frac{2\pi}{iqr} \left(e^{iqr} - e^{-iqr} \right) = 4\pi \frac{\sin(qr)}{qr}$$

---

[3] We now use $\vec{r}$ as the integration variable.

Hence

$$f_{BA}(k,\theta) = -\int_0^\infty r^2 dr\, j_0(qr) v(r)$$

Now suppose one has a Yukawa potential of the form

$$v(r) = \lambda \frac{e^{-\mu r}}{r} \qquad ; \text{ Yukawa potential}$$

The Born approximation is then given by

$$\begin{aligned}
f_{BA}(k,\theta) &= -\lambda \int_0^\infty r^2 dr \, \frac{e^{-\mu r}}{r} \frac{\sin(qr)}{qr} \\
&= -\frac{\lambda}{2iq} \int_0^\infty dr\, e^{-\mu r} \left(e^{iqr} - e^{-iqr} \right) \\
&= -\frac{\lambda}{2iq} \left(\frac{1}{\mu - iq} - \frac{1}{\mu + iq} \right)
\end{aligned}$$

Hence the Born approximation for the scattering amplitude from the Yukawa potential is given by

$$f_{BA}(k,\theta) = -\frac{\lambda}{q^2 + \mu^2}$$

We give a plot of the Born approximation cross-section for the Yukawa potential as a function of q/μ in Fig. 4.1 below.

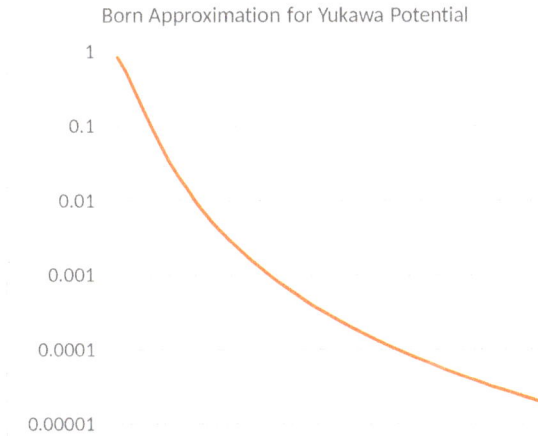

Fig. 4.1 The quantity $(\mu^4/\lambda^2)|f_{BA}|^2$ for the Yukawa potential, on a logarithmic scale, as a function of q/μ between $(0, 15)$.

Problem 4.6 Suppose one has a spherical potential of the form

$$v(r) = v_0 \qquad ; r < d$$

$$; \text{ spherical potential}$$

Show that Born approximation for the scattering amplitude is given by

$$f_{BA}(k, \theta) = -(v_0 d^3) \frac{j_1(qd)}{qd}$$

Sketch and discuss.

Solution to Problem 4.6 In this case, use the result in the previous problem

$$f_{BA}(k, \theta) = -v_0 \int_0^d r^2 dr \, j_0(qr).$$

The integral is evaluated as

$$f_{BA}(k, \theta) = -\frac{v_0}{q} \int_0^d r dr \, \sin(qr)$$

$$= \frac{v_0}{q} \frac{\partial}{\partial q} \int_0^d dr \, \cos(qr) = \frac{v_0}{q} \frac{\partial}{\partial q} \left[\frac{\sin(qd)}{q} \right]$$

$$= \frac{v_0}{q} \left[\frac{d \cos(qd)}{q} - \frac{\sin(qd)}{q^2} \right]$$

$$= -(v_0 d^3) \left[\frac{\sin(qd)}{(qd)^3} - \frac{\cos(qd)}{(qd)^2} \right]$$

Hence the Born approximation for the scattering amplitude from the spherical potential is given by

$$f_{BA}(k, \theta) = -(v_0 d^3) \frac{j_1(qd)}{qd}$$

This holds for either sign of v_0. We give a plot of the Born approximation cross-section for the spherical potential as a function of qd in Fig. 4.2 below.

If we compare the Born approximation results for the Yukawa and spherical potentials, we see that while the former continues to fall off smoothly as a function of momentum transfer, the latter exhibits a nice diffraction structure. This makes it very easy to distinguish a microscopic structure with a macroscopic laboratory scattering experiment.

The diffraction structure in Fig. 4.2 below is plotted as a function of qd where d measures the size of the potential. In order to see this structure as

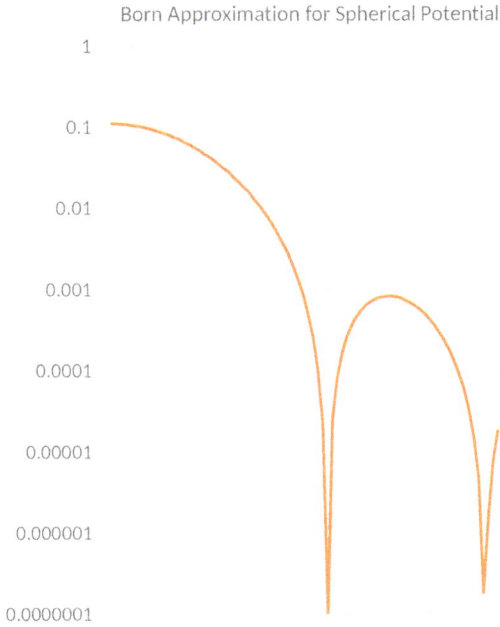

Fig. 4.2 The quantity $|f_{BA}|^2/(v_0 d^3)^2$ for the spherical potential, on a logarithmic scale, as a function of qd between $(0, 8)$.

d gets smaller and smaller, one needs higher and higher momentum transfer q. It is ironic that to see tinier and tinier structure, one needs larger and larger machines!

Chapter 5

Transition Rate

Problem 5.1 Suppose the interaction in our model problem is an integrable short-range potential of the form

$$H'(x_2, x_1) = V(|x_2 - x_1|) \qquad ; \text{ integrable potential}$$

(a) Define the momentum transfer by

$$q \equiv k_0 - k \qquad ; \text{ momentum transfer}$$

Show the matrix element of the interaction in Eqs. (5.37) takes the form

$$\langle k, n_2 | H' | k_0, n_2^0 \rangle = \tilde{V}(q) \, \tilde{\rho}_{fi}(q)$$
$$\tilde{V}(q) = \int_{-\infty}^{\infty} dx \, e^{iqx} \, V(|x|)$$
$$\tilde{\rho}_{fi}(q) = \int_0^{L_2} dx \, e^{iqx} \, \psi_{n_2}^*(x) \psi_{n_2^0}(x)$$

(b) Show the rate in Eqs. (5.37) becomes

$$\frac{1}{I_{\text{inc}}} R_{fi} \, dn_f = \left(\frac{k}{k_0} \right) \left| \left(\frac{m_1}{\hbar^2 k} \right) \tilde{V}(q) \right|^2 |\tilde{\rho}_{fi}(q)|^2$$

(c) Show this has the correct dimensions.

Solution to Problem 5.1 A summary of the results for the model one-dimensional inelastic-scattering problem developed in the text, incorporating the footnote at the bottom of page 40 there, is given in

Eqs. (5.37)

$$\frac{1}{I_{\text{inc}}} R_{fi}\, dn_f = \left(\frac{k}{k_0}\right)\left(\frac{m_1}{\hbar^2 k}\right)^2 |\langle k, n_2|H'|k_0, n_2^0\rangle|^2$$

$$\langle k, n_2|H'|k_0, n_2^0\rangle = \int_{-L_1/2}^{L_1/2} dx_1\, e^{i(k_0-k)x_1} \int_0^{L_2} dx_2\, \psi_{n_2}^*(x_2) H'(x_2, x_1)\psi_{n_2^0}(x_2)$$

Suppose the interaction in our model problem is an integrable short-range potential of the form

$$H'(x_2, x_1) = V(|x_2 - x_1|) \qquad ; \text{ integrable potential}$$

(a) The momentum transfer is defined by

$$q \equiv k_0 - k \qquad ; \text{ momentum transfer}$$

Consider the integral over x_1 in the matrix element of the interaction at a given x_2

$$\int_{-L_1/2}^{L_1/2} dx_1\, e^{iqx_1} V(|x_2 - x_1|) = e^{iqx_2} \int_{-L_1/2-x_2}^{L_1/2-x_2} dy\, e^{iqy}\, V(|y|)\, ; \, y \equiv x_1 - x_2$$

If the integral is well-convergent, the limit $L_1 \to \infty$ will not affect the answer, and

$$\int_{-L_1/2}^{L_1/2} dx_1\, e^{iqx_1} V(|x_2 - x_1|) = e^{iqx_2} \int_{-\infty}^{\infty} dy\, e^{iqy}\, V(|y|)$$

The above expression for the matrix element of the interaction then *factors* and takes the form

$$\langle k, n_2|H'|k_0, n_2^0\rangle = \tilde{V}(q)\, \tilde{\rho}_{fi}(q)$$

$$\tilde{V}(q) = \int_{-\infty}^{\infty} dy\, e^{iqy}\, V(|y|)$$

$$\tilde{\rho}_{fi}(q) = \int_0^{L_2} dx\, e^{iqx}\, \psi_{n_2}^*(x)\psi_{n_2^0}(x)$$

This is known as a *convolution*.

(b) The above expression for the rate divided by the flux then immediately becomes

$$\frac{1}{I_{\text{inc}}} R_{fi}\, dn_f = \left(\frac{k}{k_0}\right) \left|\left(\frac{m_1}{\hbar^2 k}\right) \tilde{V}(q)\right|^2 |\tilde{\rho}_{fi}(q)|^2$$

(c) This expression is evidently *dimensionless*.

Problem 5.2 The previous problem provides an expression for the ratio of the transition rate to the incident flux in our model problem of the scattering of one particle from another trapped inside a box in one dimension. It remains to evaluate the target transition matrix element $\tilde{\rho}_{fi}(q)$. Since everything now concerns the particle in the box, we can drop the superfluous subscript 2. For the transition from the ground state with $n = 1$ to an excited state with n, one needs the integral

$$\tilde{\rho}_{n,1}(q) = \frac{2}{L} \int_0^L dx\, e^{iqx} \sin\left(\frac{n\pi x}{L}\right) \sin\left(\frac{\pi x}{L}\right)$$

where the quantum number n and the size L now refer to the box.[1]

(a) Evaluate this integral and show

$$\tilde{\rho}_{n,1}(q) = \frac{1}{i} \frac{4n\pi^2(qL)}{[(qL)^2 - (n^2+1)\pi^2]^2 - (2n\pi^2)^2} \left[1 + (-1)^n e^{iqL}\right]$$

(b) Make a plot $|\tilde{\rho}_{n,1}(q)|^2$ as a function of qL. Discuss.[2]

Solution to Problem 5.2 (a) With a re-expression of the sines in terms of exponentials, one has

$$\tilde{\rho}_{n,1}(q) = \frac{2}{L} \int_0^L dx\, e^{iqx} \sin\left(\frac{n\pi x}{L}\right) \sin\left(\frac{\pi x}{L}\right)$$
$$= -\frac{1}{2L} \int_0^L dx\, e^{iqx} \left(e^{in\pi x/L} - e^{-in\pi x/L}\right)\left(e^{i\pi x/L} - e^{-i\pi x/L}\right)$$

After doing the integrals, all the numerators will be of the form

$$e^{iqL \pm in\pi \pm i\pi} - 1 = -\left[1 + (-1)^n e^{iqL}\right]$$

Hence this comes out as a common factor. The integration then gives

$$\tilde{\rho}_{n,1}(q) = \frac{1}{2i}\left[1 + (-1)^n e^{iqL}\right]\left[\frac{1}{qL + (n+1)\pi} + \frac{1}{qL - (n+1)\pi}\right.$$
$$\left. - \frac{1}{qL + (n-1)\pi} - \frac{1}{qL - (n-1)\pi}\right]$$
$$= \frac{1}{2i}\left[1 + (-1)^n e^{iqL}\right]\left[\frac{-2\pi}{(qL + n\pi)^2 - \pi^2} + \frac{2\pi}{(qL - n\pi)^2 - \pi^2}\right]$$

---

[1] Recall Fig. 3.2 in the text. Note that this integral is well-defined for all values of qL.
[2] It is of interest to make at least one log plot to see the diffraction structure.

This becomes

$$\tilde{\rho}_{n,1}(q) = \frac{\pi}{i} \left[1 + (-1)^n e^{iqL} \right] \left[\frac{1}{(qL)^2 + (n\pi)^2 - \pi^2 - 2n\pi(qL)} \right. $$
$$\left. - \frac{1}{(qL)^2 + (n\pi)^2 - \pi^2 + 2n\pi(qL)} \right]$$

which is

$$\tilde{\rho}_{n,1}(q) = \frac{1}{i} \left[1 + (-1)^n e^{iqL} \right] \frac{4n\pi^2(qL)}{\left[(qL)^2 + (n\pi)^2 - \pi^2 \right]^2 - (2n\pi)^2(qL)^2}$$

After completing the square in the variable $(qL)^2 + f(n)$, the denominator of the fraction can be rewritten as

$$D = [(qL)^2 - (n^2 + 1)\pi^2]^2 - (2n\pi^2)^2$$

Hence, one finally arrives at

$$\tilde{\rho}_{n,1}(q) = \frac{1}{i} \frac{4n\pi^2(qL)}{[(qL)^2 - (n^2 + 1)\pi^2]^2 - (2n\pi^2)^2} \left[1 + (-1)^n e^{iqL} \right]$$

(b) A plot $|\tilde{\rho}_{n,1}(q)|^2$ as a function of qL is given in Fig. 5.1 below.

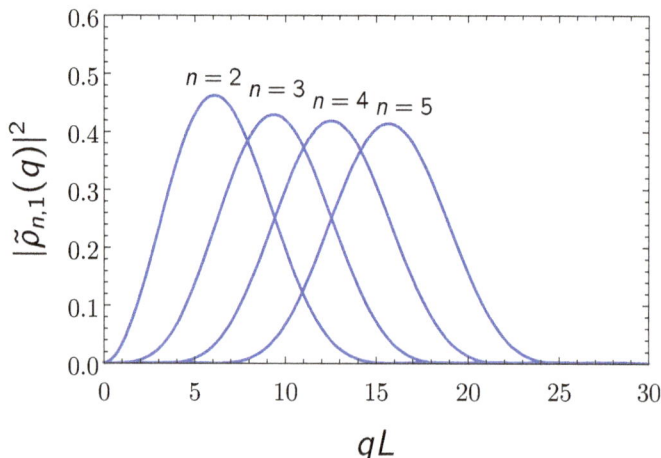

Fig. 5.1   Plot of $|\tilde{\rho}_{n,1}(q)|^2$ as a function of (qL) for four values of n. Here $L \equiv L_2$ is the size of the box. (The author would like to thank Paolo Amore for preparing this figure.)

Problem 5.3 Consider the *back scattering* of the projectile in our model problem.

(a) Show that the only modification of the results in Probs. 5.1 and 5.2 is that the momentum transfer now becomes

$$q = k_0 + k \qquad ; \text{ back scattering}$$

(b) Show that this can be made arbitrarily large for a given energy transfer, and hence one can experimentally map out the results in Prob. 5.2(b).

Solution to Problem 5.3 If we analyze *back scattering* in our model problem, then the final projectile momentum points in the negative-x direction and the final projectile wave function becomes

$$\psi_{n_1}(x) = \frac{1}{\sqrt{L}} e^{-ikx} \qquad ; \; k = \frac{2\pi n_1}{L}$$
$$; n_1 = 0, 1, 2, \cdots$$

where k is a positive number.

(a) It follows that the only modification of the results in Probs. 5.1 and 5.2 is that the momentum transfer now becomes

$$q = k_0 + k \qquad ; \text{ back scattering}$$

(b) This quantity can be made arbitrarily large for a given energy transfer, and hence one can experimentally map out the results in Prob. 5.2(b).

Problem 5.4 The goal of the two-state mixing example is to treat the interaction H' to all orders. The system starts in the state ψ_1 at $t = 0$. Expand the expression for the probability of finding the system in the state ψ_2 after the time T in Eq. (5.57), and divide by T, to get the transition rate as $H' \to 0$. Show

$$R_{21}(T) = \frac{1}{T} P_{21}(T) = \frac{1}{\hbar^2} |\langle \psi_2 | H' | \psi_1 \rangle|^2 T \qquad ; \; H' \to 0$$

Show this reproduces the *lowest-order* expression for the transition rate in Eq. (5.21) at energy conservation $\hbar\omega = 0$.

Solution to Problem 5.4 The two-state mixing problem is discussed in Sec. 5.7. The goal is to focus on the transitions in a two-state system, and treat the interaction H' exactly, instead of just in lowest order. Under the assumptions stated there, the probability of finding the system in the state ψ_2 after the time T if it started in the state ψ_1 is given in Eq. (5.57)

$$P_2(T) = |a_2(T)|^2 = \sin^2(H'_{12}\,T/\hbar)$$

Now expand this expression, and divide by T, to get the transition rate as $H' \to 0$

$$R_{fi}(T) = \frac{1}{T}P_{fi}(T) = \frac{1}{\hbar^2}|\langle\psi_2|H'|\psi_1\rangle|^2\,T \qquad ; \; H' \to 0$$

This is precisely the *lowest-order* expression for the transition rate that we calculated in Eq. (5.21) at energy conservation $\hbar\omega = 0$, which applies here since the states (ψ_1, ψ_2) are degenerate.

Chapter 6

Quantum Electrodynamics

Problem 6.1 The electromagnetic fields $(\vec{E}, \vec{B})$ are related to the potentials by

$$\vec{B}(\vec{x}, t) = \vec{\nabla} \times \vec{A}(\vec{x}, t)$$
$$\vec{E}(\vec{x}, t) = -\frac{\partial \vec{A}(\vec{x}, t)}{\partial t} - \vec{\nabla}\Phi(\vec{x}, t)$$

Let $\Lambda(\vec{x}, t)$ be a well-defined function of position and time.

(a) Show the electromagnetic fields are unchanged under a *gauge transformation*

$$\vec{A}(\vec{x}, t) \to \vec{A}(\vec{x}, t) + \vec{\nabla}\Lambda(\vec{x}, t) \qquad ; \text{ gauge transformation}$$
$$\Phi(\vec{x}, t) \to \Phi(\vec{x}, t) - \frac{\partial \Lambda(\vec{x}, t)}{\partial t}$$

Since the fields are unchanged, the physics should be unchanged.

(b) Show that the terms in Λ can be eliminated from the Schrödinger equation by making a *local phase transformation* on the wave function

$$\Psi(\vec{x}, t) \to e^{ie\Lambda(\vec{x}, t)/\hbar}\,\Psi(\vec{x}, t)$$

Remember that the wave function is *not* a physical observable.

Solution to Problem 6.1 (a) If we make the above gauge transformation on the potentials, the fields are changed to

$$\vec{B}(\vec{x}, t) \to \vec{\nabla} \times \vec{A}(\vec{x}, t) + \vec{\nabla} \times \vec{\nabla}\Lambda(\vec{x}, t)$$
$$\vec{E}(\vec{x}, t) \to -\frac{\partial \vec{A}(\vec{x}, t)}{\partial t} - \frac{\partial}{\partial t}\vec{\nabla}\Lambda(\vec{x}, t) - \vec{\nabla}\Phi(\vec{x}, t) + \vec{\nabla}\frac{\partial \Lambda(\vec{x}, t)}{\partial t}$$

Now observe the following:

(1) It is a general property of vectors that

$$\vec{\nabla} \times \vec{\nabla} = 0$$

(2) The order of the partial derivatives can always be interchanged, so that

$$\vec{\nabla} \frac{\partial}{\partial t} = \frac{\partial}{\partial t} \vec{\nabla}$$

Hence under the above gauge transformation the physical fields are *unchanged*

$$\vec{B}(\vec{x},t) \to \vec{B}(\vec{x},t)$$
$$\vec{E}(\vec{x},t) \to \vec{E}(\vec{x},t)$$

(b) If the combination $(\vec{p} - e\vec{A})$ appears in the hamiltonian, then the local phase change under the gauge transformation passes right through this combination

$$(\vec{p} - e\vec{A})\Psi \to e^{ie\Lambda(\vec{x},t)/\hbar} \left[\frac{\hbar}{i}\vec{\nabla} + e\vec{\nabla}\Lambda(\vec{x},t) - e\vec{A}(\vec{x},t) - e\vec{\nabla}\Lambda(\vec{x},t) \right] \Psi(\vec{x},t)$$

$$= e^{ie\Lambda(\vec{x},t)/\hbar}(\vec{p} - e\vec{A})\Psi$$

If $e\Phi$ appears linearly in the hamiltonian, then the local phase change similarly passes right through the combination $[i\hbar(\partial/\partial t) - e\Phi]$

$$\left(i\hbar\frac{\partial}{\partial t} - e\Phi \right) \Psi(\vec{x},t) \to e^{ie\Lambda(\vec{x},t)/\hbar} \times$$

$$\left[i\hbar\frac{\partial}{\partial t} - e\frac{\partial\Lambda(\vec{x},t)}{\partial t} - e\Phi(\vec{x},t) + e\frac{\partial\Lambda(\vec{x},t)}{\partial t} \right] \Psi(\vec{x},t)$$

$$= e^{ie\Lambda(\vec{x},t)/\hbar} \left(i\hbar\frac{\partial}{\partial t} - e\Phi \right) \Psi(\vec{x},t)$$

This local phase then moves through all the operators in the Schrödinger equation

$$\left(i\hbar\frac{\partial}{\partial t} - H \right) \Psi \to e^{ie\Lambda(\vec{x},t)/\hbar} \left(i\hbar\frac{\partial}{\partial t} - H \right) \Psi$$

where it can just be cancelled. The Schrödinger equation is therefore *invariant* under a gauge transformation coupled together with a local phase transformation on the wave function

$$\left(i\hbar \frac{\partial}{\partial t} - H \right) \Psi = 0 \qquad ; \text{ Schrödinger equation}$$

Problem 6.2 Suppose one has chosen a vector potential $\vec{A}'(\vec{x}, t)$ which reproduces the $(\vec{E}, \vec{B})$ fields, and suppose the *divergence* of this field is non-zero

$$\vec{\nabla} \cdot \vec{A}'(\vec{x}, t) = \rho'(\vec{x}, t)$$

(a) Make a gauge transformation as in Prob. 6.1, and show that with the new vector potential $\vec{A}(\vec{x}, t)$

$$\vec{\nabla} \cdot \vec{A}(\vec{x}, t) = \rho'(\vec{x}, t) + \nabla^2 \Lambda(\vec{x}, t)$$

(b) Conclude that one can always work in the *Coulomb gauge*, where

$$\vec{\nabla} \cdot \vec{A}(\vec{x}, t) = 0 \qquad ; \text{ Coulomb gauge}$$

Solution to Problem 6.2 (a) Suppose one has an initial set of potentials $(\vec{A}', \Phi')$ with

$$\vec{\nabla} \cdot \vec{A}'(\vec{x}, t) = \rho'(\vec{x}, t)$$

Make the gauge transformation of the previous problem to a new set of potentials $(\vec{A}, \Phi)$, in which case

$$\vec{\nabla} \cdot \vec{A}(\vec{x}, t) = \rho'(\vec{x}, t) + \nabla^2 \Lambda(\vec{x}, t)$$

The inhomogeneous Laplace's equation can always be spatially integrated at a given time to find a $\Lambda(\vec{x}, t)$ such that

$$\nabla^2 \Lambda(\vec{x}, t) = -\rho'(\vec{x}, t)$$

Hence

$$\vec{\nabla} \cdot \vec{A}(\vec{x}, t) = 0$$

(b) In terms of the potentials $(\vec{A}, \Phi)$, the electric field is given by

$$\vec{E}(\vec{x}, t) = -\frac{\partial \vec{A}(\vec{x}, t)}{\partial t} - \vec{\nabla} \Phi(\vec{x}, t)$$

Take the divergence of this relation, and use the fourth of Maxwell's equations

$$\vec{\nabla} \cdot \vec{E}(\vec{x}, t) = -\nabla^2 \Phi(\vec{x}, t) = \frac{\rho(\vec{x}, t)}{\varepsilon_0}$$

where $\rho(\vec{x}, t)$ is the *charge density*. This equation can similarly be spatially integrated at a given time to find the potential $\Phi(\vec{x}, t)$.

Hence one can always work in the Coulomb gauge with potentials $(\vec{A}, \Phi)$ *satisfying*

$$\vec{\nabla} \cdot \vec{A}(\vec{x}, t) = 0 \qquad \text{; Coulomb gauge}$$

Problem 6.3 Show that if one were to retain just the first term in the time-dependent interaction in Eq. (6.17), so that

$$H' = -e\mathcal{E}_0 x \cos(\omega_0 t) \doteq -\left(\frac{e\mathcal{E}_0}{2}\right) x\, e^{i\omega_0 t}$$

then the rate R_{fi} in Eq. (6.20) would become

$$R_{fi} = \left(\frac{e\mathcal{E}_0}{2}\right)^2 \frac{2\pi}{\hbar} |\langle f|x|i\rangle|^2 \delta(E_f - E_i + \hbar\omega_0) \qquad \text{; de-excitation}$$

Compare this with the rate R_{fi} in Eq. (6.20)

$$R_{fi} = \left(\frac{e\mathcal{E}_0}{2}\right)^2 \frac{2\pi}{\hbar} |\langle f|x|i\rangle|^2 \delta(E_f - E_i - \hbar\omega_0) \qquad \text{; excitation}$$

Now fix E_i and $\hbar\omega_0$, and conclude the following:

- As a function of E_f there are two non-overlapping rate expressions, one for $E_f = E_i - \hbar\omega_0$, and one for $E_f = E_i + \hbar\omega_0$;
- Since there is no overlap of the two expressions, one can ignore any *interference term* between them in the calculation of the rates;
- In leading order, the real time-dependent perturbation H' in Eq. (6.17) therefore describes both *de-excitation* and *excitation* of the system;
- Correspondingly, the perturbation H' in Eq. (6.26) describes both *emission* and *absorption* of radiation.

Solution to Problem 6.3 In the text, we show what happens to the transition rate for ionization by an oscillating electric field if we retain only the *final* term in the interaction in Eq. (6.17), which pumps energy into the system

$$H' = -e\mathcal{E}_0 x \cos(\omega_0 t) \doteq -\left(\frac{e\mathcal{E}_0}{2}\right) x \, e^{-i\omega_0 t}$$

The resulting transition rate in displayed in Eq. (6.20)

$$R_{fi} = \left(\frac{e\mathcal{E}_0}{2}\right)^2 \frac{2\pi}{\hbar} |\langle f|x|i\rangle|^2 \delta(E_f - E_i - \hbar\omega_0) \qquad ; \text{ excitation}$$

Suppose that one were to retain just the *first* term in the time-dependent interaction in Eq. (6.17)

$$H' = -e\mathcal{E}_0 x \cos(\omega_0 t) \doteq -\left(\frac{e\mathcal{E}_0}{2}\right) x \, e^{i\omega_0 t}$$

The calculation of the rate goes through again exactly as in the text, and the result in Eq. (6.20) becomes

$$R_{fi} = \left(\frac{e\mathcal{E}_0}{2}\right)^2 \frac{2\pi}{\hbar} |\langle f|x|i\rangle|^2 \delta(E_f - E_i + \hbar\omega_0) \qquad ; \text{ de-excitation}$$

The only change is in the argument of the energy-conserving delta-function.

As stated in the problem, we now fix E_i and $\hbar\omega_0$, and conclude the following:

- As a function of E_f there are two non-overlapping rate expressions, one for $E_f = E_i - \hbar\omega_0$, and one for $E_f = E_i + \hbar\omega_0$;
- Since there is no overlap of the two expressions, one can ignore any *interference term* between them in the calculation of the rates;
- In leading order, the real time-dependent perturbation H' in Eq. (6.17) therefore describes both *de-excitation* and *excitation* of the system;
- Correspondingly, the perturbation H' in Eq. (6.26) describes both *emission* and *absorption* of radiation.

Problem 6.4 This problem reviews the main aspects of *adjoints* and *hermiticity*. For simplicity, we go back to one dimension and express the matrix element of an operator N between two acceptable wave functions as

$$\langle \psi_f | N | \psi_i \rangle = \int dx \, \psi_f^*(x) N \psi_i(x)$$

The *adjoint* operator $N^\dagger$ is then defined through the relation

$$\int dx \ \left[N^\dagger\psi_f(x)\right]^* \psi_i(x) = \int dx \ \psi_f^*(x)N\psi_i(x) \qquad ; \text{ adjoint}$$

An operator is *hermitian* if it is identical to its adjoint

$$N^\dagger = N \qquad ; \text{ hermitian}$$

(a) The eigenstates of N satisfy

$$N\psi_n(x) = n\,\psi_n(x)$$

Take the matrix element of this relation with $\psi_n^*(x)$, and prove that the eigenvalues of a hermitian operator are *real*;

(b) Suppose one has a *pair* of operators MN. Show

$$(MN)^\dagger = N^\dagger M^\dagger$$

(c) Re-verify that the operators $p = (\hbar/i)d/dx$ and $H = p^2/2m + V(x)$ with a real $V(x)$ are hermitian;

(d) Show that in two-dimensions, the angular momentum operator L_z is hermitian.

Solution to Problem 6.4 (a) Since the operator N is hermitian, it follows that

$$\langle\psi_n|N|\psi_n\rangle^* = \langle\psi_n|N|\psi_n\rangle$$

The matrix element of the eigenvalue equation gives

$$\langle\psi_n|N|\psi_n\rangle = n\langle\psi_n|\psi_n\rangle = n$$

We conclude that the eigenvalue is *real*

$$n = n^*$$

(b) The adjoint statement for the operator pair MN is

$$\int dx \ \left[(MN)^\dagger\psi_f(x)\right]^* \psi_i(x) = \int dx \ \psi_f^*(x)MN\psi_i(x)$$

Since $N\psi_i(x)$ is just another wave function, the r.h.s. can be written

$$\int dx \ \psi_f^*(x)MN\psi_i(x) = \int dx \ [M^\dagger\psi_f(x)]^*N\psi_i(x)$$

Since $M^\dagger \psi_f(x)$ is similarly just another wave function, this can be written

$$\int dx \, [M^\dagger \psi_f(x)]^\star N \psi_i(x) = \int dx \, [N^\dagger M^\dagger \psi_f(x)]^\star \psi_i(x)$$

We conclude that

$$\int dx \, \left[(MN)^\dagger \psi_f(x)\right]^\star \psi_i(x) = \int dx \, [N^\dagger M^\dagger \psi_f(x)]^\star \psi_i(x)$$

Hence

$$(MN)^\dagger = N^\dagger M^\dagger$$

The order of the operators is reversed when the adjoint is taken.

(c) The fact that the operators $p = (\hbar/i)d/dx$ and $H = p^2/2m + V(x)$ with a real $V(x)$ are hermitian is demonstrated in Secs. 2.5 and 3.1 in the text. Since these arguments are central to the development in the text, readers are asked to work through them again for themselves.

(d) In two-dimensions, the angular momentum operator L_z is given by

$$L_z = \frac{\hbar}{i} \left(x \frac{\partial}{\partial y} - y \frac{\partial}{\partial x} \right)$$

Consider the matrix element

$$\langle \psi_f | L_z | \psi_i \rangle = \frac{\hbar}{i} \int dx \, dy \, \psi_f^\star(x, y) \left(x \frac{\partial}{\partial y} - y \frac{\partial}{\partial x} \right) \psi_i(x, y)$$

Suppose the boundary conditions are such that the system is *localized*, so that the wave functions vanish when the coordinates go off to infinity in either direction. Then the boundary contributions vanish when the first term is partially integrated on y at fixed x, and when the second term is partially integrated on x at fixed y. We conclude that

$$\langle \psi_f | L_z | \psi_i \rangle = \langle \psi_i | L_z | \psi_f \rangle^\star$$

With these boundary conditions, the operator L_z is hermitian.[1]

Problem 6.5 (a) Show that the number eigenstate $|n\rangle$ in Sec. 6.7 can actually be constructed as

$$|n\rangle = \frac{1}{\sqrt{n!}} (a^\dagger)^n |0\rangle$$

---

[1]Can you find other boundary conditions that leave L_z hermitian?

  (b) Make use of the commutation relations to show this state is normalized

$$\langle n|n \rangle = 1$$

Solution to Problem 6.5

(a) From Eq. (6.68), one has for the simple harmonic oscillator

$$a^\dagger |n\rangle = \sqrt{n+1}\, |n+1\rangle$$

Hence the normalized state with $n+1$ quanta can be obtained from that with n quanta by

$$|n+1\rangle = \frac{1}{\sqrt{n+1}} a^\dagger |n\rangle$$

If we start with the state $|0\rangle$ of zero quanta, we can the build up the state $|n\rangle$ by repeated application of this relation

$$|n\rangle = \frac{1}{\sqrt{n}} a^\dagger \frac{1}{\sqrt{n-1}} a^\dagger \cdots \frac{1}{\sqrt{2}} a^\dagger \frac{1}{\sqrt{1}} a^\dagger |0\rangle$$

Hence

$$|n\rangle = \frac{1}{\sqrt{n!}} (a^\dagger)^n |0\rangle$$

  (b) The norm of this state is then

$$\langle n|n \rangle = \frac{1}{n!} \langle 0| a^n (a^\dagger)^n |0\rangle$$

Take the first a over to the right using the identity

$$a(a^\dagger)^n \equiv [a, a^\dagger](a^\dagger)^{n-1} + a^\dagger [a, a^\dagger](a^\dagger)^{n-2} + \cdots + (a^\dagger)^{n-1}[a, a^\dagger] + (a^\dagger)^n a$$

Now make use of the relations

$$[a, a^\dagger] = 1 \qquad\qquad ; \ a|0\rangle = 0$$

so that

$$a(a^\dagger)^n |0\rangle = n(a^\dagger)^{n-1} |0\rangle$$

It follows that

$$\langle n|n \rangle = \frac{1}{(n-1)!} \langle 0| a^{n-1} (a^\dagger)^{n-1} |0\rangle$$

Repetition of this reduction n times gives

$$\langle n|n \rangle = \langle 0|0 \rangle = 1$$

which demonstrates that the state is indeed normalized.

Problem 6.6 Repeat the calculation of photoionization starting from the quantized radiation field and a single photon in the state $|1_{\vec{k}s}\rangle$. Use an incident photon flux of

$$I_{\text{inc}} = \frac{c}{\Omega} \qquad ; \text{ photon flux}$$

Show the *photoionization cross-section* becomes[2]

$$\sigma_{fi} \equiv \frac{1}{I_{\text{inc}}} R_{fi}\, dn_f \qquad ; \text{ photoionization}$$

$$= \frac{\alpha}{2\pi c^2}\left(\frac{mc}{\hbar}\right)\left(\frac{k_f}{k}\right)\left|\vec{e}_{\vec{k}s} \cdot \int d^3 x\, e^{i(\vec{k}-\vec{k}_f)\cdot\vec{x}}\left(\frac{\vec{p}}{m}\right)\psi_0(\vec{x})\right|^2 d\Omega_f$$

Solution to Problem 6.6 Photoionization is discussed in Sec. 6.3. With the quantized radiation field, the initial photon state has one photon in the state $|1_{\vec{k}s}\rangle$ and the vector potential in Eq. (6.77) then makes the transition to the state $|0\rangle$. Hence the rate expression in Eq. (6.30) now takes the form

$$\frac{1}{I_{\text{inc}}} R_{fi}\, dn_f = e^2 \left(\frac{\Omega}{c}\right)\left(\frac{\hbar}{2\varepsilon_0 \omega_k \Omega}\right)\left(\frac{1}{m^2 L^3}\right) \times$$

$$\left(\frac{2\pi}{\hbar}\right)\left|\vec{e}_{\vec{k}s} \cdot \int d^3 x\, e^{i(\vec{k}-\vec{k}_f)\cdot\vec{x}}\,\vec{p}\,\psi_0(\vec{x})\right|^2 \delta(E_f - E_0 - \hbar\omega)\left[\frac{L^3}{(2\pi)^3} d^3 k_f\right]$$

The rest of the calculation proceeds exactly as in Sec. 6.3 to give

$$\sigma_{fi} \equiv \frac{1}{I_{\text{inc}}} R_{fi}\, dn_f \qquad ; \text{ photoionization}$$

$$= \frac{\alpha}{2\pi c^2}\left(\frac{mc}{\hbar}\right)\left(\frac{k_f}{k}\right)\left|\vec{e}_{\vec{k}s} \cdot \int d^3 x\, e^{i(\vec{k}-\vec{k}_f)\cdot\vec{x}}\left(\frac{\vec{p}}{m}\right)\psi_0(\vec{x})\right|^2 d\Omega_f$$

This is also a very nice result, since it is the photoionization cross-section for any quantum system, albeit with a plane-wave particle in the final state.

Problem 6.7 The required transition matrix element for the quantum system in the general expression for the photon emission rate in Eq. (6.81)

---

[2] In the text, the polarization unit vector $\hat{e}_{\vec{k},s}$ should be written as $\vec{e}_{\vec{k},s}$.

is

$$\vec{M}_{fi} = \int d^3x \, e^{-i\vec{k}\cdot\vec{x}} \, \psi_f^*(\vec{x}) \, \frac{\vec{p}}{m} \, \psi_i(\vec{x})$$

(a) Suppose the hamiltonian for that quantum system has the form

$$H = \frac{\vec{p}^2}{2m} + V(\vec{x})$$

Show

$$[H, \vec{x}] = \frac{\hbar}{i} \frac{\vec{p}}{m}$$

Hence, show

$$\vec{M}_{fi} = \frac{i}{\hbar} \int d^3x \, e^{-i\vec{k}\cdot\vec{x}} \, \psi_f^*(\vec{x}) \, [H, \vec{x}] \, \psi_i(\vec{x})$$

(b) Now assume the wavelength of light is such that $kR \ll 1$ where R is a measure of the size of the system. Then

$$e^{-i\vec{k}\cdot\vec{x}} \approx 1 \qquad\qquad ; \, kR \ll 1$$

Show that in this limit the required matrix element becomes

$$\vec{M}_{fi} = \frac{i}{\hbar}(E_f - E_i) \int d^3x \, \psi_f^*(\vec{x}) \, \vec{x} \, \psi_i(\vec{x}) \quad ; \, \text{dipole approximation}$$

This is known as the *dipole approximation.*

Solution to Problem 6.7 (a) Make use of the relation

$$[p_j p_j, x_k] = p_j[p_j, \, x_k] + [p_j, \, x_k]p_j$$
$$= 2p_k\frac{\hbar}{i}$$

It follows that

$$[H, \vec{x}] = \frac{1}{2m}[\vec{p}^2, \, \vec{x}] = \frac{\hbar}{i} \frac{\vec{p}}{m}$$

Hence

$$\vec{M}_{fi} = \frac{i}{\hbar} \int d^3x \, e^{-i\vec{k}\cdot\vec{x}} \, \psi_f^*(\vec{x}) \, [H, \vec{x}] \, \psi_i(\vec{x})$$

(b) Now assume the wavelength of the light is such that $kR \ll 1$ where R is a measure of the size of the system. Then in the integral

$$e^{-i\vec{k}\cdot\vec{x}} \approx 1 \qquad\qquad ; \, kR \ll 1$$

The states in the integral are eigenstates of energy so that

$$H\psi_i(\vec{x}) = E_i\psi_i(\vec{x}) \qquad ; \; H\psi_f(\vec{x}) = E_f\psi_f(\vec{x})$$

Therefore[3]

$$\vec{M}_{fi} = \frac{i}{\hbar}(E_f - E_i) \int d^3x \; \psi_f^*(\vec{x}) \, \vec{x} \, \psi_i(\vec{x}) \quad ; \text{ dipole approximation}$$

This is known as the *dipole approximation*.

---

[3] We use the fact that the hamiltonian is hermitian and the eigenvalues are real.

Chapter 7

Quantum Statistics

Problem 7.1 Consider a non-interacting spin-1/2 Fermi gas inside a big box of volume $V = L^3$ with p.b.c. in all three directions. In its ground state, the levels are filled up to a wavenumber $|\vec{k}| = k_F$.

(a) Count the number of filled levels. Show the number of particles per unit volume is

$$n_0 \equiv \frac{N}{V} = \frac{1}{V} \left[\frac{2L^3}{(2\pi)^3} \int_0^{k_F} d^3k \right] = \frac{k_F^3}{3\pi^2}$$

(b) Show the energy per particle is

$$\frac{E}{N} = \frac{3}{5}\varepsilon_F \qquad ; \varepsilon_F = \frac{(\hbar k_F)^2}{2m}$$

where ε_F is the *Fermi energy*.

(c) Compute the pressure from the first law of thermodynamics $P = -(dE/dV)$.[1] Show

$$P = \frac{2}{5}\frac{\hbar^2}{2m}(3\pi^2)^{2/3} n_0^{5/3} \qquad ; \text{ Fermi gas}$$

Solution to Problem 7.1 The Pauli Exclusion Principle states that you can only put one identical fermion into any given state. Consider a non-interacting spin-1/2 Fermi gas inside a big box of volume $V = L^3$ with p.b.c. in all three directions. In its ground state, the levels are filled up to a wavenumber $|\vec{k}| = k_F$, and we can put two fermions, one with spin up and one with spin down, into any given spatial state.

---

[1] Here the many-body system is in its ground state at a temperature $T = 0$.

(a) The number of states in the interval d^3k in this problem is given by

$$d^3n = \frac{2V}{(2\pi)^3} d^3k \qquad ; V = L^3$$

The number of particles per unit volume in the filled Fermi sea follows as

$$n_0 \equiv \frac{N}{V} = \left[\frac{2}{(2\pi)^3} \int_0^{k_F} d^3k \right] = \frac{k_F^3}{3\pi^2}$$

(b) The energy per particle is then given by

$$\frac{E}{N} = \frac{\hbar^2}{2m} \frac{1}{(N/V)} \left[\frac{2}{(2\pi)^3} \int_0^{k_F} k^2 d^3k \right] = \frac{3}{5}\varepsilon_F \qquad ; \varepsilon_F = \frac{(\hbar k_F)^2}{2m}$$

where ε_F is the *Fermi energy*.

(c) The pressure can be computed from the first law of thermodynamics, $P = -(dE/dV)$, for a system in its ground state at $T = 0$

$$P = -N\frac{d}{dV} \left\{ \frac{3}{5} \frac{\hbar^2}{2m} \left[3\pi^2 \left(\frac{N}{V}\right) \right]^{2/3} \right\}$$

$$= \frac{2}{5} \frac{\hbar^2}{2m} (3\pi^2)^{2/3} n_0^{5/3} \qquad ; \text{Fermi gas}$$

These are very valuable relations, and it is particularly important to realize that because of all the momenta involved, a Fermi gas in its ground state exerts a *pressure depending on its density*.

Problem 7.2 Consider a non-interacting spin-0 Bose gas of massive particles inside a big box of volume $V = L^3$ with p.b.c. in all three directions. In its ground state, the particles all occupy the $\vec{k} = 0$ level.

(a) Compute the pressure from the first law of thermodynamics $P = -(dE/dV)$. Show

$$P = 0 \qquad ; \text{Bose gas}$$

(b) Suppose instead, that the particles are confined to a large cubicle box of volume $V = L^3$, where the ground-state single-particle energy is $\varepsilon_0 = (\hbar^2/2m)(3\pi^2/L^2)$. Show the pressure is

$$P = \frac{2}{3}\varepsilon_0 n_0 \qquad ; n_0 = \frac{N}{V}$$

Compare with the result in part (a). Discuss.

Solution to Problem 7.2 Suppose that instead of the Fermi gas in Problem 7.1, we have a non-interacting spin-0 Bose gas of massive particles inside a big box of volume $V = L^3$ with p.b.c. in all three directions. In its ground state, the particles all occupy the $\vec{k} = 0$ level.

(a) The ground-state energy is $E_0 = 0$, and if we again compute the pressure $P = -(dE/dV)$ from the first law of thermodynamics for a system in its ground state at $T = 0$, we obtain

$$P = -\frac{dE_0}{dV} = 0 \qquad ; \text{ Bose gas}$$

The ground-state wave function is constant, there is no curvature, and there is no kinetic energy.

(b) Suppose instead, that the particles are confined to a large cubicle box of volume $V = L^3$, where the ground-state single-particle energy is

$$\varepsilon_0 = \frac{\hbar^2}{2m}\left(\frac{3\pi^2}{L^2}\right) = \frac{\hbar^2}{2m}\left(\frac{3\pi^2}{V^{2/3}}\right)$$

The ground-state energy of the Bose system is then

$$E_0 = N\frac{\hbar^2}{2m}\left(\frac{3\pi^2}{V^{2/3}}\right)$$

The pressure follows as is

$$P = \frac{2}{3}\varepsilon_0 n_0 \qquad ; n_0 = \frac{N}{V}$$

Now there is a small pressure since the particles are confined to the box and there is curvature in the wave function. At any particle density n_0, however, this pressure vanishes as $V \to \infty$.

Problem 7.3 While the full quantum many-body problem is complicated, the Hartree self-consistent field approximation provides a surprisingly good first approximation for many finite systems such as the electron cloud in the atom or the atomic nucleus. Suppose there is a two-body interaction $V(|\vec{x} - \vec{y}|)$. The Hartree single-particle potential is defined by

$$U_H(\vec{x}) = \int d^3y\, V(|\vec{x} - \vec{y}|)n(\vec{y})$$

where the particle density is given by the sum over all the occupied states of the absolute square of the single-particle wave functions

$$n(\vec{y}) = \sum_i |\psi_i(\vec{y})|^2 \qquad ; \text{ over occupied states}$$

These wave functions, in turn, are given by the solutions in the self-consistent potential

$$\left[-\frac{\hbar^2\nabla^2}{2m} + U_H(\vec{x})\right]\psi_i(\vec{x}) = \varepsilon_i\psi_i(\vec{x})$$

Provide a physical motivation for the Hartree equations, and give some indication as to how you would go about solving them.

Solution to Problem 7.3 The first approximation to a particle moving through a many-body medium with a density $n(\vec{x})$ is to say there is some corresponding average potential $U(\vec{x})$ in which the particle moves. If there is a two-particle potential $V(|\vec{x}-\vec{y}|)$ between the particles, then the first approximation to this average potential is just

$$U(\vec{x}) = \int d^3y\, V(|\vec{x}-\vec{y}|)n(\vec{y})$$

If the wave function for the many-body system looks like an occupied set of single-particle orbitals (a major assumption!), then one is led to the Hartree expression for the density, and a corresponding set of self-consistent Hartree single-particle equations

$$n_H(\vec{y}) = \sum_i |\psi_i(\vec{y})|^2 \qquad ; \text{ over occupied states}$$

$$\left[-\frac{\hbar^2\nabla^2}{2m} + U_H(\vec{x})\right]\psi_i(\vec{x}) = \varepsilon_i\psi_i(\vec{x})$$

These equations can be solved by *iteration*:

- Assume a form for the density;
- Solve the Hartee equations numerically for the wave functions;
- Recompute the density;
- Resolve for the wave functions;
- Hope that this process converges!

Problem 7.4 Consider a Bose gas where essentially all the particles remain in the condensate. Suppose the two-particle interaction is a contact interaction.

(a) Show that the Hartree potential takes the form

$$U_H(\vec{x}) = \lambda|\phi_0(\vec{x})|^2$$

where ϕ_0 is the condensate wave function in Eqs. (7.16) and (7.17), and λ is a constant independent of N.

(b) Show that the Hartree equation for that condensate wave function then becomes

$$\left[-\frac{\hbar^2\nabla^2}{2m} + \lambda|\phi_0(\vec{x})|^2\right]\phi_0(\vec{x}) = \varepsilon_0\phi_0(\vec{x})$$

Note that this is now simply a local, nonlinear, differential equation.

Solution to Problem 7.4 Bose condensation is discused in Sec. 7.1.1. The condensate wave function $\phi_0(\vec{x})$ is related to the single-particle ground-state wave function $\psi_0(\vec{x})$ in Eq. (7.16)[2]

$$\sqrt{N}\,\psi_0(\vec{x}) = \phi_0(\vec{x})$$

The particle density is

$$n(\vec{x}) = N|\psi_0(\vec{x})|^2 = |\phi_0(\vec{x})|^2$$

We are given that the interparticle potential is a contact interaction of the form

$$V(\vec{x}-\vec{y}) = \lambda\,\delta^{(3)}(\vec{x}-\vec{y})$$

(a) The Hartree single-particle potential is defined in the previous problem

$$U_H(\vec{x}) = \int d^3y\, V(|\vec{x}-\vec{y}|)n(\vec{y})$$

where $n(\vec{y})$ is the particle density. Substitution of the above relations gives

$$U_H(\vec{x}) = \lambda\int d^3y\,\delta^{(3)}(\vec{x}-\vec{y})\,n(\vec{y}) = \lambda\,n(\vec{x})$$
$$= \lambda\,|\phi_0(\vec{x})|^2$$

(b) Now multiply the Hartree single-particle equation in the previous problem by $\sqrt{N}$

$$\left[-\frac{\hbar^2\nabla^2}{2m} + U_H(\vec{x})\right]\phi_0(\vec{x}) = \varepsilon_0\phi_0(\vec{x})$$

This is now simply a *local, nonlinear, differential equation* for the condensate wave function.

---

[2] We have set the phase $\xi_0 = 1$.

Problem 7.5 Consider the non-interacting Bose condensate wave function $\phi_0 = \sqrt{N}\psi_0$ in Eq. (7.16), where ψ_0 is the ground-state single-particle level.[3] Suppose the particles are in a big box with p.b.c in all directions.

(a) Show the square of the modulus is

$$|\phi_0|^2 = n_0 \qquad \text{; particle density}$$

where n_0 is the particle density;

(b) Compute the probability flux $\vec{S}$ from Eq. (4.28) using this wave function, and show the medium is at rest

$$\vec{S}(\phi_0) = 0$$

(c) The medium can be given a velocity by including a spatially-dependent phase in the single-particle wave function $\psi_0 \to \psi_0 \, e^{i\chi(\vec{x})}$. Show the expression for the particle density is unchanged

$$|\phi_0 e^{i\chi}|^2 = n_0$$

(d) The velocity of the medium can now be identified from the probability flux calculated with this new wave function. Give an argument that

$$n_0\vec{v} = \vec{S}\left(\phi_0 e^{i\chi}\right)$$

Hence obtain

$$\vec{v} = \frac{\hbar}{m}\vec{\nabla}\chi \qquad \text{; fluid velocity}$$

The fluid velocity of the Bose condensate is obtained from the gradient of the phase of the single-particle wave function.

(e) Show the fluid motion is irrotational

$$\vec{\nabla} \times \vec{v} = 0 \qquad \text{; irrotational}$$

Solution to Problem 7.5 (a) In the non-interacting Bose condensate, there are N particles occupying the lowest energy single-particle eigenstate $\psi_0(\vec{x})$. Hence the density is

$$n_0 = N|\psi_0(\vec{x})|^2 = |\phi_0(\vec{x})|^2 \qquad \text{; } \phi_0 = \sqrt{N}\psi_0$$

Although, as stated, this problem is for a constant density n_0, the results are more general.

---

[3]We choose $\xi = 1$.

(b) The ground-state wave function ψ_0 is real, and hence the probability current in Eq. (4.28) vanishes

$$\vec{S}(\phi_0) = \frac{\hbar}{2im}\left[\phi_0^\star \vec{\nabla}\phi_0 - \left(\vec{\nabla}\phi_0\right)^\star \phi_0\right] = 0$$

(c) The medium can now be given a velocity by including a spatially-dependent phase in the single-particle wave function $\psi_0 \to \psi_0\, e^{i\chi(\vec{x})}$. The expression for the particle density is evidently unchanged

$$|\phi_0 e^{i\chi}|^2 = n_0$$

The probability flux computed with this new wave function is

$$\vec{S}\left(\phi_0 e^{i\chi}\right) = \frac{\hbar}{2im}\left\{\left[\phi_0\, e^{i\chi(\vec{x})}\right]^\star \vec{\nabla}\left[\phi_0\, e^{i\chi(\vec{x})}\right] - \left[\vec{\nabla}\phi_0\, e^{i\chi(\vec{x})}\right]^\star \left[\phi_0\, e^{i\chi(\vec{x})}\right]\right\}$$

It is only the gradient of the exponential that gives a non-zero value, and

$$\vec{S}\left(\phi_0 e^{i\chi}\right) = \frac{\hbar}{m}\vec{\nabla}\chi\,|\phi_0|^2$$

(d) We can now say that the whole medium has been put into motion and, just as in Prob. 2.5, we identify the velocity of the medium from

$$\vec{S}\left(\phi_0 e^{i\chi}\right) = \vec{v}\,|\phi_0|^2 = \vec{v}\,n_0$$

Hence, we obtain

$$\vec{v}(\vec{x}) = \frac{\hbar}{m}\vec{\nabla}\chi(\vec{x}) \qquad ;\ \text{fluid velocity}$$

The fluid velocity of the Bose condensate is obtained from the gradient of the phase of the single-particle wave function.

This is a remarkable quantum-mechanical relation between a macroscopic observable and the properties of a single-particle wave function.

(e) Since the curl of the gradient vanishes, the fluid motion is irrotational

$$\vec{\nabla} \times \vec{v} = 0 \qquad ;\ \text{irrotational}$$

Problem 7.6 Repeat Prob. 7.1 for non-interacting spin-1/2 particles in a one-dimensional box.

(a) Show the particle density is given by

$$n_0 \equiv \frac{N}{L} = \frac{1}{L}\left[\frac{2L}{\pi}\int_0^{k_F} dk\right] = \frac{2k_F}{\pi}$$

(b) Show the energy per particle is

$$\frac{E}{N} = \frac{1}{3}\varepsilon_F \qquad\qquad ; \; \varepsilon_F = \frac{(\hbar k_F)^2}{2m}$$

(c) Show these are the *same results* one gets for particles moving on the large circle with p.b.c. (Remember to include both directions!)

Solution to Problem 7.6 We can repeat Problem 7.1 for a one-dimensional Fermi gas of spin-1/2 fermions confined to a box of length L. Here the number of states is related to the interval dk by

$$dn = \frac{2L}{\pi} dk$$

where there are two fermions in each state.

(a) The particle density, the number of particles per unit length, is then given by

$$n_0 \equiv \frac{N}{L} = \left[\frac{2}{\pi} \int_0^{k_F} dk \right] = \frac{2k_F}{\pi}$$

(b) The energy per particle follows as

$$\frac{E}{N} = \frac{\hbar^2}{2m} \frac{1}{(N/L)} \left[\frac{2}{\pi} \int_0^{k_F} k^2 \, dk \right] = \frac{1}{3}\varepsilon_F \qquad ; \; \varepsilon_F = \frac{(\hbar k_F)^2}{2m}$$

(c) It is interesting to note that if, instead of being confined to the box the particles move on a length L with p.b.c., one obtains *exactly the same results*. Here the number of states in the interval dk about k is now

$$dn = \frac{2L}{2\pi} dk$$

However, one has to include states moving in both directions so that in (a), for example, one gets the same answer

$$n_0 \equiv \frac{N}{L} = \left[\frac{2}{2\pi} \int_{-k_F}^{k_F} dk \right] = \frac{2k_F}{\pi}$$

The same holds true for E/N.

Problem 7.7 One can go from plane-polarized photons to photons with a given helicity through a *canonical transformation*. Define

$$\vec{e}_{\vec{k},\pm 1} \equiv \mp \frac{1}{\sqrt{2}} \left(\vec{e}_{\vec{k}1} \pm i\vec{e}_{\vec{k}2} \right) \qquad ; \; b_{\vec{k},\pm 1} \equiv \mp \frac{1}{\sqrt{2}} \left(b_{\vec{k}1} \mp ib_{\vec{k}2} \right)$$

(a) Show

$$\sum_{s=1,2} \vec{e}_{\vec{k}s} b_{\vec{k}s} = \sum_{\lambda=\pm 1} \vec{e}_{\vec{k},\lambda} b_{\vec{k},\lambda} \qquad ; \qquad \sum_{s=1,2} \vec{e}_{\vec{k}s} b^{\dagger}_{\vec{k}s} = \sum_{\lambda=\pm 1} \vec{e}^{\dagger}_{\vec{k},\lambda} b^{\dagger}_{\vec{k},\lambda}$$

(b) Show

$$[b_{\vec{k},\lambda}, b^{\dagger}_{\vec{k},\lambda'}] = \delta_{\lambda,\lambda'} \qquad ; \ [b_{\vec{k},\lambda}, b_{\vec{k},\lambda'}] = [b^{\dagger}_{\vec{k},\lambda}, b^{\dagger}_{\vec{k},\lambda'}] = 0$$

The properties of the operators follow from these commutation relations.

Solution to Problem 7.7 Define the transformation from plane polarization to helicity by

$$\vec{e}_{\vec{k},\pm 1} \equiv \mp \frac{1}{\sqrt{2}} \left(\vec{e}_{\vec{k}1} \pm i\vec{e}_{\vec{k}2} \right) \qquad ; \ b_{\vec{k},\pm 1} \equiv \mp \frac{1}{\sqrt{2}} \left(b_{\vec{k}1} \mp i b_{\vec{k}2} \right)$$

(a) Compute the following expression

$$\sum_{\lambda=\pm 1} \vec{e}_{\vec{k},\lambda} b_{\vec{k},\lambda} = \frac{1}{2} \left(\vec{e}_{\vec{k}1} + i\vec{e}_{\vec{k}2} \right) \left(b_{\vec{k}1} - i b_{\vec{k}2} \right) + \frac{1}{2} \left(\vec{e}_{\vec{k}1} - i\vec{e}_{\vec{k}2} \right) \left(b_{\vec{k}1} + i b_{\vec{k}2} \right)$$

$$= \sum_{s=1,2} \vec{e}_{\vec{k}s} b_{\vec{k}s}$$

Also compute

$$\sum_{\lambda=\pm 1} \vec{e}^{\dagger}_{\vec{k},\lambda} b^{\dagger}_{\vec{k},\lambda} = \frac{1}{2} \left(\vec{e}_{\vec{k}1} - i\vec{e}_{\vec{k}2} \right) \left(b^{\dagger}_{\vec{k}1} + i b^{\dagger}_{\vec{k}2} \right) + \frac{1}{2} \left(\vec{e}_{\vec{k}1} + i\vec{e}_{\vec{k}2} \right) \left(b^{\dagger}_{\vec{k}1} - i b^{\dagger}_{\vec{k}2} \right)$$

$$= \sum_{s=1,2} \vec{e}_{\vec{k}s} b^{\dagger}_{\vec{k}s}$$

These are the quantities that appear in the vector potential field operator in quantum electrodynamics.

(b) Since the destruction and creation operators commute among themselves

$$[b_{\vec{k},\lambda}, b_{\vec{k},\lambda'}] = [b^{\dagger}_{\vec{k},\lambda}, b^{\dagger}_{\vec{k},\lambda'}] = 0,$$

now compute

$$[b_{\vec{k},+1}, b^{\dagger}_{\vec{k},+1}] = \frac{1}{2}[(b_{\vec{k}1} - i b_{\vec{k}2}), (b^{\dagger}_{\vec{k}1} + i b^{\dagger}_{\vec{k}2})] = 1$$

$$[b_{\vec{k},-1}, b^{\dagger}_{\vec{k},-1}] = \frac{1}{2}[(b_{\vec{k}1} + i b_{\vec{k}2}), (b^{\dagger}_{\vec{k}1} - i b^{\dagger}_{\vec{k}2})] = 1.$$

Also

$$[b_{\vec{k},+1}, b^\dagger_{\vec{k},-1}] = -\frac{1}{2}[(b_{\vec{k}1} - ib_{\vec{k}2}), (b^\dagger_{\vec{k}1} - ib^\dagger_{\vec{k}2})] = 0$$

$$[b_{\vec{k},-1}, b^\dagger_{\vec{k},+1}] = -\frac{1}{2}[(b_{\vec{k}1} + ib_{\vec{k}2}), (b^\dagger_{\vec{k}1} + ib^\dagger_{\vec{k}2})] = 0$$

Hence, in summary, for the helicity creation and destruction operators

$$[b_{\vec{k},\lambda}, b^\dagger_{\vec{k},\lambda'}] = \delta_{\lambda,\lambda'}$$

We have shown that all of the properties of these operators follow from the commutation relations.

Chapter 8

Quantum Measurements

Problem 8.1 This problem discusses the *spin* of spin-1/2 fermions.[1] The main thing to keep in mind here is that the spin operators and wave functions are very *simple*. The spin operator is

$$\hbar \vec{S} = \frac{\hbar}{2}\vec{\sigma} \qquad ; \text{ spin operator}$$

where $\vec{\sigma}$ are the 2×2 Pauli matrices $(\sigma_x, \sigma_y, \sigma_z)$ defined in Eqs. (11.54). The spin wave functions ("spinors") for spin up and down along the z-axis are

$$\phi_\uparrow = \begin{pmatrix} 1 \\ 0 \end{pmatrix} \qquad ; \phi_\downarrow = \begin{pmatrix} 0 \\ 1 \end{pmatrix}$$

(a) Show these are eigenstates of σ_z

$$\sigma_z \phi_\uparrow = \phi_\uparrow \qquad ; \sigma_z \phi_\downarrow = -\phi_\downarrow$$

(b) The eigenstates of spin up and down along the x-axis can be constructed as

$$\phi_\rightarrow = \frac{1}{\sqrt{2}}(\phi_\uparrow + \phi_\downarrow) \qquad ; \phi_\leftarrow = \frac{1}{\sqrt{2}}(\phi_\uparrow - \phi_\downarrow)$$

Show

$$\sigma_x \phi_\rightarrow = \phi_\rightarrow \qquad ; \sigma_x \phi_\leftarrow = -\phi_\leftarrow$$

(c) Hence, show

$$\phi_\uparrow = \frac{1}{\sqrt{2}}(\phi_\rightarrow + \phi_\leftarrow) \qquad ; \phi_\downarrow = \frac{1}{\sqrt{2}}(\phi_\rightarrow - \phi_\leftarrow)$$

---

[1]This problem involves simple matrix manipulations. If these are unfamiliar to you, please do Probs. 11.5 and 11.6 first.

Solution to Problem 8.1 The *Pauli matrices* are given by

$$\sigma_x = \begin{pmatrix} 0 & 1 \\ 1 & 0 \end{pmatrix} \quad ; \; \sigma_y = \begin{pmatrix} 0 & -i \\ i & 0 \end{pmatrix} \quad ; \; \sigma_z = \begin{pmatrix} 1 & 0 \\ 0 & -1 \end{pmatrix} \quad ; \; \text{Pauli matrices}$$

The spin wave functions ("spinors") for spin up and down along the z-axis are

$$\phi_\uparrow = \begin{pmatrix} 1 \\ 0 \end{pmatrix} \qquad ; \; \phi_\downarrow = \begin{pmatrix} 0 \\ 1 \end{pmatrix}$$

(a) It follows that these are eigenstates of σ_z

$$\sigma_z \phi_\uparrow = \begin{pmatrix} 1 & 0 \\ 0 & -1 \end{pmatrix} \begin{pmatrix} 1 \\ 0 \end{pmatrix} = \begin{pmatrix} 1 \\ 0 \end{pmatrix} = \phi_\uparrow$$

$$\sigma_z \phi_\downarrow = \begin{pmatrix} 1 & 0 \\ 0 & -1 \end{pmatrix} \begin{pmatrix} 0 \\ 1 \end{pmatrix} = \begin{pmatrix} 0 \\ -1 \end{pmatrix} = -\phi_\downarrow$$

(b) In a similar manner, one has

$$\sigma_x \phi_\uparrow = \begin{pmatrix} 0 & 1 \\ 1 & 0 \end{pmatrix} \begin{pmatrix} 1 \\ 0 \end{pmatrix} = \begin{pmatrix} 0 \\ 1 \end{pmatrix} = \phi_\downarrow$$

$$\sigma_x \phi_\downarrow = \begin{pmatrix} 0 & 1 \\ 1 & 0 \end{pmatrix} \begin{pmatrix} 0 \\ 1 \end{pmatrix} = \begin{pmatrix} 1 \\ 0 \end{pmatrix} = \phi_\uparrow$$

The Pauli matrix σ_x just interchanges these eigenstates. The eigenstates of spin up and down along the x-axis are then immediately constructed as

$$\phi_\rightarrow = \frac{1}{\sqrt{2}} (\phi_\uparrow + \phi_\downarrow) \qquad ; \; \phi_\leftarrow = \frac{1}{\sqrt{2}} (\phi_\uparrow - \phi_\downarrow)$$

for clearly

$$\sigma_x \phi_\rightarrow = \phi_\rightarrow \qquad ; \; \sigma_x \phi_\leftarrow = -\phi_\leftarrow$$

(c) The above relations are immediately inverted to express the eigenstates along the z-axis in terms of those along the x-axis

$$\phi_\uparrow = \frac{1}{\sqrt{2}} (\phi_\rightarrow + \phi_\leftarrow) \qquad ; \; \phi_\downarrow = \frac{1}{\sqrt{2}} (\phi_\rightarrow - \phi_\leftarrow)$$

Problem 8.2 The Stern-Gerlach detector in Fig. 8.1 in the text is rotated by 90° about the y-axis, and the beam with spin-up along the x-axis is separated. The detector is now returned to its initial configuration in Fig. 8.1 and that beam is inserted into it. Express the state $| \rightarrow \rangle$ with spin-up along the x-axis as a linear combination of the states $|\uparrow\rangle$ and $|\downarrow\rangle$ with spin-up and spin-down along the z-axis, and describe what is now observed coming out of that detector returned to its initial configuration.

Solution to Problem 8.2 It follows from part (b) of the previous problem that for the one-particle states in the abstract occupation number space

$$| \rightarrow \rangle = \frac{1}{\sqrt{2}} |\uparrow\rangle + \frac{1}{\sqrt{2}} |\downarrow\rangle$$

Thus the beam now coming out of detector returned to its initial configuration consists of both spin-up and spin-down along the z-axis with equal probability amplitudes.

The intermediate measurement of spin along the x-axis has produced a new state involving both components of spin along the z-axis.

Problem 8.3 (a) Consider a particle of mass m_0 moving in a circle of radius a in the (x, y)-plane, as discussed in Probs. 2.2, 2.4 and 2.5. Suppose the particle has a charge e. We know from E&M that the little current loop has a magnetic moment $\mu_z = (\pi a^2) i_e$, where i_e is the current. Show

$$i_e = e\nu = e\frac{\omega}{2\pi} = e\frac{L_z}{2\pi I}$$

Hence, show the magnetic moment is

$$\mu_z = \frac{e}{2m_0} L_z$$

(b) When the angular momentum is quantized with $L_z = \hbar m$, show

$$\mu_z = \mu m \qquad ; \mu = \frac{e\hbar}{2m_0}$$
$$; m = 0, \pm 1, \pm 2, \cdots$$

(c) Discuss how the Stern-Gerlach apparatus can be used to prepare a state $\psi_m(\phi) = e^{im\phi}/\sqrt{2\pi}$;

(d) Discuss how photoabsorption on the ground state can be used to prepare a state $\psi = a\psi_m(\phi) + b\psi_{-m}(\phi)$.

Solution to Problem 8.3 (a) We know from E&M that the magnetic field of a little current loop in the (x, y)-plane looks just like that of a magnetic dipole of moment[2]

$$\mu_z = (\pi a^2)i$$

Here πa^2 is the area of the little circular loop, and i is the current — the charge passing a given point per unit time. If the frequency of the circular motion is ν, then the current is

$$i = e\nu = e\frac{\omega}{2\pi}$$

In addition, for circular motion in the (x, y)-plane, the z-component of the angular momentum is

$$L_z = m_0 a\nu = m_0 a^2 \omega$$

Hence

$$\mu_z = \frac{e}{2m_0}L_z$$

(b) It was shown in Prob. 2.2 that the angular momentum L_z is quantized according to

$$L_z = \hbar m \qquad ; m = 0, \pm 1, \pm 2, \cdots$$

Hence the magnetic moment μ_z is given by

$$\mu_z = \mu m \qquad ; \mu = \frac{e\hbar}{2m_0}$$
$$; m = 0, \pm 1, \pm 2, \cdots$$

(c) Equation (8.1) now applies, and the analysis of the Stern-Gerlach experiment is exactly that in Sec. 8.1. An extracted beam in a pure pass measurement at the time t_0 has its state vector reduced to

$$|\psi_{\text{int}}(t_0)\rangle = \frac{c_m(t_0)}{|c_m(t_0)|}|m\rangle$$

with a corresponding wave function

$$\langle \phi | m \rangle = \frac{1}{\sqrt{2\pi}}e^{im\phi}$$

---

[2]See [Walecka (2018)].

(d) The energy of the particle of mass m_0 moving in a circular orbit of radius a was calculated in Prob. 2.2 to be

$$E_m = \frac{\hbar^2}{2I} m^2 \qquad ; I = m_0 a^2$$

If we control the energy of the photon in photoabsorption to go from the ground state with energy E_0 to a state with energy E_m, then we will have prepared a state of the target with wave function $\psi = a\psi_m(\phi) + b\psi_{-m}(\phi)$. The magnitude of the amplitudes in this expression are given by the transition rates from the ground state. The relative phase has to be determined through some other means.

Chapter 9

Formal Structure of Quantum Mechanics

Problem 9.1 Consider a particle moving in one dimension in a large circle of length L and satisfying periodic boundary conditions. The wave functions are[1]

$$\langle x|k\rangle = \psi_k(x) = \frac{1}{\sqrt{L}}e^{ikx}$$

$$k = \frac{2\pi n}{L} \qquad\qquad ; n = 0, \pm 1, \cdots$$

(a) Use completeness of the eigenstates of position to show that

$$\langle k|k'\rangle = \delta_{k,k'}$$

(b) Show that the matrix elements of the kinetic energy operator are

$$\langle k|\hat{T}|k'\rangle = \frac{(\hbar k)^2}{2m}\delta_{k,k'}$$

(c) Show that the matrix elements of the potential $V(\hat{x})$ are

$$\langle k|\hat{V}|k'\rangle = \frac{1}{L}\tilde{V}(k-k')$$

$$\tilde{V}(k-k') = \int_0^L dx\, V(x)\, e^{-i(k-k')x}$$

---

[1] Any piecewise continuous function $\psi(x)$ can actually be expanded in this set, and the basis functions are *complete* in the sense that

$$\text{Lim}_{N\to\infty} \int_0^L dx\, \left|\psi(x) - \sum_{n=-N}^{N} c_k\psi_k(x)\right|^2 = 0 \qquad ; \text{ completeness}$$

This is all the completeness we will need for the physics in this volume.

Solution to Problem 9.1 (a) The problem of a particle moving in a large circle of length L and satisfying p.b.c. is discussed in Sec. 2.5. The wave functions are given by

$$\langle x|k \rangle = \psi_k(x) = \frac{1}{\sqrt{L}} e^{ikx}$$

$$k = \frac{2\pi n}{L} \qquad \qquad ; \, n = 0, \pm 1, \cdots$$

They satisfy the orthonormality relation in Eq. (2.33), which here, with the use of the completeness relation for the eigenstates of position, is written as

$$\langle k|k' \rangle = \int_0^L dx \, \langle k|x \rangle \langle x|k' \rangle = \frac{1}{L} \int_0^L dx \, e^{i(k'-k)x}$$

$$= \delta_{k,k'}$$

(b) The states are eigenstates of momentum with eigenvalue $\hbar k$ [2]

$$\hat{p}|k \rangle = \hbar k|k \rangle$$

It follows that the matrix elements of the kinetic energy operator $\hat{T} = \hat{p}^2/2m$ are

$$\langle k|\hat{T}|k' \rangle = \frac{(\hbar k')^2}{2m} \langle k|k' \rangle = \frac{(\hbar k)^2}{2m} \delta_{k,k'}$$

(c) The matrix elements of the potential $V(\hat{x})$ are calculated using the completeness of the eigenstates of position (twice)

$$\langle k|\hat{V}|k' \rangle = \int dx \int dx' \, \langle k|x \rangle \langle x|V(\hat{x})|x' \rangle \langle x'|k' \rangle$$

$$= \int dx \int dx' \, V(x')\delta(x - x')\langle k|x \rangle \langle x'|k' \rangle$$

$$= \int dx \, V(x)\langle k|x \rangle \langle x|k' \rangle$$

$$= \frac{1}{L} \int_0^L dx \, V(x) \, e^{-i(k-k')x}$$

This is defined as the Fourier transform of the potential (recall Prob. 5.1)

$$\langle k|\hat{V}|k' \rangle = \frac{1}{L}\tilde{V}(k - k')$$

---

[2] The eigenvalue $\hbar k$ is the amount of momentum the particle carries while undergoing this one-dimensional motion around the large circle; for a concrete realization here, consider, for example, a bead moving without friction on a large circular wire.

Problem 9.2 Look for stationary states in the abstract Hilbert space

$$|\Psi(t)\rangle = e^{-iEt/\hbar}\,|\Psi\rangle$$
$$\hat{H}|\Psi\rangle = E|\Psi\rangle$$

Define the amplitude in the momentum representation as

$$\langle k|\Psi\rangle \equiv A(k)$$

(a) Use the results of Prob. 9.1 to show that the Schrödinger equation in the momentum representation is

$$\sum_{k'}\left\{\left[E - \frac{(\hbar k)^2}{2m}\right]\delta_{k,k'} - \frac{1}{L}\tilde{V}(k-k')\right\}A(k') = 0$$

(b) Use the density of final states in one dimension to convert this to the *integral equation*

$$\left[E - \frac{(\hbar k)^2}{2m}\right]A(k) - \frac{1}{2\pi}\int_{-\infty}^{\infty}dk'\,\tilde{V}(k-k')A(k') = 0$$

Note that the differential equation for the amplitude $\langle x|\Psi\rangle$ in the coordinate representation, and the integral equation for the amplitude $\langle k|\Psi\rangle$ in the momentum representation, both follow from the *same* Schrödinger equation in the abstract Hilbert space!

Solution to Problem 9.2 As stated, we look for stationary states in the abstract Hilbert space

$$|\Psi(t)\rangle = e^{-iEt/\hbar}\,|\Psi\rangle$$
$$\hat{H}|\Psi\rangle = E|\Psi\rangle$$

We also define the amplitude in the momentum representation as

$$\langle k|\Psi\rangle \equiv A(k)$$

(a) Now take the matrix element of the Schrödinger equation, and insert a complete set of eigenstates of momentum

$$\sum_{k'}\left[\langle k|\hat{T}|k'\rangle + \langle k|\hat{V}|k'\rangle\right]\langle k'|\Psi\rangle = E\langle k|\Psi\rangle$$

With the use of the results from the previous problem, this is rewritten as

$$\sum_{k'}\left\{\left[E - \frac{(\hbar k)^2}{2m}\right]\delta_{k,k'} - \frac{1}{L}\tilde{V}(k-k')\right\}A(k') = 0$$

(b) The density of final states in one dimension with p.b.c. and a length L is

$$dn = \frac{L}{2\pi} dk'$$

Hence we can convert the relation in part (a) to an *integral equation* for the amplitude $A(k)$

$$\left[E - \frac{(\hbar k)^2}{2m} \right] A(k) - \frac{1}{2\pi} \int_{-\infty}^{\infty} dk' \, \tilde{V}(k - k') A(k') = 0$$

As stated, we note that the differential equation for the amplitude $\langle x|\Psi \rangle$ in the coordinate representation, and the integral equation for the amplitude $\langle k|\Psi \rangle$ in the momentum representation, both follow from the *same* Schrödinger equation in the abstract Hilbert space!

Problem 9.3 The abstract states in Probs. 9.1 and 9.2 are eigenstates of the hermitian momentum operator $\hat{p}$

$$\hat{p}|k\rangle = \hbar k|k\rangle$$

where the basic commutation relation in the abstract Hilbert space is

$$[\hat{p}, \hat{x}] = \frac{\hbar}{i}$$

(a) Verify that the eigenvalues $\hbar k$ are real;

(b) Show the projection of the first relation on the eigenstate of position is

$$\langle x|\hat{p}|k\rangle = \frac{\hbar}{i} \frac{\partial}{\partial x} \langle x|k\rangle = \hbar k \langle x|k\rangle$$

(c) Show that in the momentum representation, the canonical commutation relation is satisfied with

$$\langle k|\hat{x}|k'\rangle = i \frac{\partial}{\partial k} \delta_{k,k'} \qquad ; \text{ momentum rep}$$

Solution to Problem 9.3 We use the states and wave functions of Prob. 9.1 for a particle moving in a large circle of length L and satisfying p.b.c., as discussed in Sec. 2.5. The wave functions are given by

$$\langle x|k\rangle = \psi_k(x) = \frac{1}{\sqrt{L}} e^{ikx}$$

$$k = \frac{2\pi n}{L} \qquad ; n = 0, \pm 1, \cdots$$

(a) The operator $\hat{p}$ is hermitian. Therefore

$$\langle k|\hat{p}|k\rangle = \langle k|\hat{p}|k\rangle^{\star}$$

It follows that the eigenvalues are real

$$k = k^{\star}$$

(b) Consider the matrix element

$$\langle x|\hat{p}|k\rangle = \hbar k\langle x|k\rangle$$

With the above wave functions, this can be written[3]

$$\hbar k\langle x|k\rangle = \hbar k\left(\frac{1}{\sqrt{L}}e^{ikx}\right) = \frac{\hbar}{i}\frac{d}{dx}\left(\frac{1}{\sqrt{L}}e^{ikx}\right) = \frac{\hbar}{i}\frac{d}{dx}\langle x|k\rangle$$

(c) Use the completeness of the eigenstates of position to write

$$\langle k|\hat{x}|k'\rangle = \int_0^L dx\, x\langle k|x\rangle\langle x|k'\rangle = i\frac{d}{dk}\left(\frac{1}{L}\int_0^L dx\, e^{-i(k-k')x}\right)$$

$$= i\frac{d}{dk}\delta_{k,k'}$$

It is shown in Eq. (2.11) that in coordinate space where $p = (\hbar/i)d/dx$, there is a basic *commutation relation*

$$[p,x]\psi(x) = (px - xp)\psi(x) = \frac{\hbar}{i}\psi(x)$$

Now in momentum space with momentum $p = \hbar k$, with position given by the differential operator $x = i\,d/dk$, and with a wave function $A(k)$, the *same* commutation relation is satisfied

$$[p,x]A(k) = (px - xp)A(k) = \frac{\hbar}{i}A(k)$$

Problem 9.4 The completeness relation in the abstract Hilbert space for the eigenstates of momentum for the particle moving on the large circle of length L with p.b.c. reads

$$\sum_k |k\rangle\langle k| = \hat{1} \qquad ;\text{ completeness}$$

$$k = \frac{2\pi n}{L} \qquad ;\, n = 0, \pm 1, \cdots$$

---

[3] Since there are no other space-time coordinates here, we can simply write $(\hbar/i)d/dx$, without the partial.

(a) Use this to show

$$\langle x|x'\rangle = \sum_k \langle x|k\rangle\langle k|x'\rangle = \frac{1}{L}\sum_k e^{ik(x-x')} = \delta(x-x')$$

This is the completeness relation for the complex Fourier series;

(b) Similarly, use the result in Prob. 9.1(b) to show that

$$\langle x|\hat{T}|x'\rangle = \sum_k\sum_{k'}\langle x|k\rangle\langle k|\hat{T}|k'\rangle\langle k'|x'\rangle = \sum_k \frac{(\hbar k)^2}{2m}\langle x|k\rangle\langle k|x'\rangle$$

$$= -\frac{\hbar^2}{2m}\frac{\partial^2}{\partial x^2}\langle x|x'\rangle$$

(c) Show

$$\langle x|\hat{p}|x'\rangle = \sum_k\sum_{k'}\langle x|k\rangle\langle k|\hat{p}|k'\rangle\langle k'|x'\rangle = \sum_k \hbar k\langle x|k\rangle\langle k|x'\rangle$$

$$= \frac{\hbar}{i}\frac{\partial}{\partial x}\langle x|x'\rangle$$

Solution to Problem 9.4 We are given that the completeness relation in the abstract Hilbert space for the eigenstates of momentum for the particle moving on the large circle of length L with p.b.c. reads

$$\sum_k |k\rangle\langle k| = \hat{1} \qquad ; \text{ completeness}$$

$$k = \frac{2\pi n}{L} \qquad ; n = 0,\pm 1,\cdots$$

(a) Start from the inner product $\langle x|x'\rangle$, and insert the unit operator

$$\langle x|x'\rangle = \langle x|\hat{1}|x'\rangle = \sum_k \langle x|k\rangle\langle k|x'\rangle = \frac{1}{L}\sum_k e^{ik(x-x')}$$

This reproduces the completeness relation for the complex Fourier series

$$\langle x|x'\rangle = \sum_k \langle x|k\rangle\langle k|x'\rangle = \frac{1}{L}\sum_k e^{ik(x-x')} = \delta(x-x')$$

We will find these relations extremely useful.

(b) Verify the inner product $\langle k|k'\rangle$

$$\langle k|k'\rangle = \sum_x \langle k|x\rangle\langle x|k'\rangle = \frac{1}{L}\int_0^L dx\, e^{i(k'-k)x} = \delta_{k,k'}$$

It follows that the matrix element of the kinetic energy operator in momentum space is

$$\langle k|\hat{T}|k'\rangle = \frac{(\hbar k')^2}{2m}\langle k|k'\rangle = \frac{(\hbar k)^2}{2m}\delta_{k,k'}$$

Now evaluate the matrix element of the kinetic energy operator in coordinate-space using the above completeness relation in momentum-space[4]

$$\langle x|\hat{T}|x'\rangle = \sum_k\sum_{k'}\langle x|k\rangle\langle k|\hat{T}|k'\rangle\langle k'|x'\rangle = \sum_k\frac{(\hbar k)^2}{2m}\langle x|k\rangle\langle k|x'\rangle$$

$$= -\frac{\hbar^2}{2m}\frac{\partial^2}{\partial x^2}\sum_k\langle x|k\rangle\langle k|x'\rangle = -\frac{\hbar^2}{2m}\frac{\partial^2}{\partial x^2}\langle x|x'\rangle$$

(c) We can similarly compute the matrix element of the momentum operator in coordinate space

$$\langle x|\hat{p}|x'\rangle = \sum_k\sum_{k'}\langle x|k\rangle\langle k|\hat{p}|k'\rangle\langle k'|x'\rangle = \sum_k\hbar k\,\langle x|k\rangle\langle k|x'\rangle$$

$$= \frac{\hbar}{i}\frac{\partial}{\partial x}\sum_k\langle x|k\rangle\langle k|x'\rangle = \frac{\hbar}{i}\frac{\partial}{\partial x}\langle x|x'\rangle$$

Problem 9.5 These are two problems on the *adjoint*.

(a) Show from Eq. (9.18) that

$$(i\hat{F})^\dagger = -i\hat{F}^\dagger$$

(b) Define $|\Psi_m(t)\rangle \equiv e^{i\hat{H}t/\hbar}|\psi_m\rangle$. Use the fact that $\hat{H}$ is hermitian to show that

$$\langle\Psi_m(t)|\psi_n\rangle = \langle\psi_m|^{-i\hat{H}t/\hbar}|\psi_n\rangle$$

Solution to Problem 9.5 (a) The *adjoint* operator $\hat{F}^\dagger$ is defined through Eq. (9.18)

$$\langle\psi_n|\hat{F}|\psi_m\rangle^* = \langle\psi_m|\hat{F}^\dagger|\psi_n\rangle \qquad ; \text{ adjoint}$$

where $|\psi_m\rangle$ and $|\psi_n\rangle$ are any two acceptable states in the space.

---

[4]Here the partial derivative just reminds us that in the Schrödinger equation, the other member of the pair (x,t) is to be kept fixed.

Consider the new operator $i\hat{F}$. Then

$$\langle\psi_n|i\hat{F}|\psi_m\rangle^* = -i\langle\psi_n|\hat{F}|\psi_m\rangle^* = -i\langle\psi_m|\hat{F}^\dagger|\psi_n\rangle = \langle\psi_m| - i\hat{F}^\dagger|\psi_n\rangle$$
$$= \langle\psi_m|(i\hat{F})^\dagger|\psi_n\rangle$$

It follows that

$$(i\hat{F})^\dagger = -i\hat{F}^\dagger$$

(b) An operator is *hermitian* if it is self-adjoint

$$\hat{F} = \hat{F}^\dagger \qquad\qquad ; \text{ hermitian}$$

Consider the state

$$|\Psi_m(t)\rangle \equiv e^{i\hat{H}t/\hbar}|\psi_m\rangle = \left(1 + \frac{i}{\hbar}\hat{H}t + \cdots\right)|\psi_m\rangle$$

The fact that $\hat{H}$ is hermitian, and use of the result in part (a), gives

$$\langle\Psi_m(t)|\psi_n\rangle = \langle\psi_n|\Psi_m(t)\rangle^* = \langle\psi_n|\left(1 + \frac{i}{\hbar}\hat{H}t + \cdots\right)|\psi_m\rangle^*$$

$$= \langle\psi_m|\left(1 - \frac{i}{\hbar}\hat{H}t + \cdots\right)|\psi_n\rangle$$

$$= \langle\psi_m|^{-i\hat{H}t/\hbar}|\psi_n\rangle$$

Problem 9.6 The hamiltonian and number operator for the simple harmonic oscillator are given by

$$\hat{H}_0 = \hbar\omega\left(\hat{N} + \frac{1}{2}\right) \qquad\qquad ; \hat{N} = a^\dagger a$$

(a) Expand the exponential, rearrange the terms, and show that

$$e^{i\hat{H}_0t/\hbar}\, a\, e^{-i\hat{H}_0t/\hbar} = 1 + \frac{it}{\hbar}[\hat{H}_0, a] + \frac{1}{2!}\left(\frac{it}{\hbar}\right)^2[\hat{H}_0, [\hat{H}_0, a]] + \cdots$$

(b) Use the commutation relations for the harmonic oscillator, and show that through this order

$$e^{i\hat{H}_0t/\hbar}\, a\, e^{-i\hat{H}_0t/\hbar} = a\, e^{-i\omega t}$$

(c) Similarly, show

$$e^{i\hat{H}_0t/\hbar}\, a^\dagger\, e^{-i\hat{H}_0t/\hbar} = a^\dagger\, e^{i\omega t}$$

Solution to Problem 9.6 (a) Expansion of the exponentials gives

$$
e^{i\hat{H}_0 t/\hbar}\, a\, e^{-i\hat{H}_0 t/\hbar} = \left[1 + \left(\frac{it}{\hbar}\right)\hat{H}_0 + \frac{1}{2!}\left(\frac{it}{\hbar}\right)^2 \hat{H}_0^2 + \cdots \right] a \times
$$
$$
\left[1 - \left(\frac{it}{\hbar}\right)\hat{H}_0 + \frac{1}{2!}\left(\frac{it}{\hbar}\right)^2 \hat{H}_0^2 + \cdots \right]
$$

This can be rearranged to read

$$
e^{i\hat{H}_0 t/\hbar}\, a\, e^{-i\hat{H}_0 t/\hbar} = a + \frac{it}{\hbar}\left(\hat{H}_0 a - a\hat{H}_0 \right) +
$$
$$
\frac{1}{2!}\left(\frac{it}{\hbar}\right)^2 \left(\hat{H}_0^2 a - 2\hat{H}_0 a\hat{H}_0 + a\hat{H}_0^2 \right) + \cdots
$$

which is the same as

$$
e^{i\hat{H}_0 t/\hbar}\, a\, e^{-i\hat{H}_0 t/\hbar} = a + \frac{it}{\hbar}[\hat{H}_0, a] + \frac{1}{2!}\left(\frac{it}{\hbar}\right)^2 [\hat{H}_0, [\hat{H}_0, a]] + \cdots
$$

(b) The commutation relations for the harmonic oscillator give us

$$
[\hat{H}_0, a] = \hbar\omega[a^\dagger a, a] = \hbar\omega(a^\dagger a a - a a^\dagger a) = \hbar\omega\,[a^\dagger, a]a = -\hbar\omega a
$$

Hence

$$
e^{i\hat{H}_0 t/\hbar}\, a\, e^{-i\hat{H}_0 t/\hbar} = a + (-i\omega t)a + \frac{1}{2!}(-i\omega t)^2 a + \cdots = a\, e^{-i\omega t}
$$

(c) The only difference for $a^\dagger$ is the commutation relation

$$
[\hat{H}_0, a^\dagger] = \hbar\omega[a^\dagger a, a^\dagger] = \hbar\omega(a^\dagger a a^\dagger - a^\dagger a^\dagger a) = \hbar\omega\, a^\dagger[a, a^\dagger] = \hbar\omega a^\dagger
$$

Hence

$$
e^{i\hat{H}_0 t/\hbar}\, a^\dagger\, e^{-i\hat{H}_0 t/\hbar} = a^\dagger + (i\omega t)a^\dagger + \frac{1}{2!}(i\omega t)^2 a^\dagger + \cdots = a^\dagger\, e^{i\omega t}
$$

Problem 9.7 Consider the time development operator in the interaction picture in Eq. (9.45)

$$
\hat{U}(t, t_0) = \hat{1} - \frac{i}{\hbar}\int_{t_0}^{t} dt'\, \hat{H}_1(t') + \left(-\frac{i}{\hbar}\right)^2 \int_{t_0}^{t} dt' \int_{t_0}^{t'} dt''\, \hat{H}_1(t')\hat{H}_1(t'') + \cdots
$$

Show that the double integral in the third term can be rewritten as

$$\int_{t_0}^{t} dt' \int_{t_0}^{t'} dt'' \, \hat{H}_1(t')\hat{H}_1(t'') = \frac{1}{2!} \int_{t_0}^{t} dt' \int_{t_0}^{t} dt'' \, T[\hat{H}_1(t')\hat{H}_1(t'')]$$

where the *time-ordering* operation $T[\hat{H}_1(t')\hat{H}_1(t'')]$ places the operator with the latest time to the left.

Solution to Problem 9.7 Consider the double integral

$$\frac{1}{2!} \int_{t_0}^{t} dt' \int_{t_0}^{t} dt'' \, T[\hat{H}_1(t')\hat{H}_1(t'')]$$

This goes over a square of size $(t - t_0)$ in the (t', t'')-plane. A diagonal out from (t_0, t_0) converts this square into two triangular regions — one with $t' > t''$, and one with $t'' > t'$. The double integral, with the time-ordered integrand, can then be written as the sum of two terms

$$\frac{1}{2!} \int_{t_0}^{t} dt' \int_{t_0}^{t} dt'' \, T[\hat{H}_1(t')\hat{H}_1(t'')] = \frac{1}{2!} \int_{t_0}^{t} dt' \int_{t_0}^{t'} dt'' \, \hat{H}_1(t')\hat{H}_1(t'') +$$

$$\frac{1}{2!} \int_{t_0}^{t} dt'' \int_{t_0}^{t''} dt' \, \hat{H}_1(t'')\hat{H}_1(t')$$

A simple swap of dummy variables shows the two terms on the r.h.s. are identical, and therefore

$$\int_{t_0}^{t} dt' \int_{t_0}^{t'} dt'' \, \hat{H}_1(t')\hat{H}_1(t'') = \frac{1}{2!} \int_{t_0}^{t} dt' \int_{t_0}^{t} dt'' \, T[\hat{H}_1(t')\hat{H}_1(t'')]$$

Chapter 10

Quantum Mechanics Postulates

Problem 10.1 The statement of completeness for the eigenstates of the hermitian operator $\hat{F}$ in abstract Hilbert space is given in Eq. (9.23)[1]

$$|\Psi\rangle = \sum_f c_f |\psi_f\rangle \qquad ; \text{ completeness}$$

(a) Obtain the expansion coefficient as

$$c_f = \langle\psi_f|\Psi\rangle$$

(b) Substitute this back in to get

$$|\Psi\rangle = \sum_f |\psi_f\rangle\langle\psi_f|\Psi\rangle$$

(c) Conclude that the statement of completeness for the eigenstates of the hermitian operator $\hat{F}$ in abstract Hilbert space can be written as

$$\sum_f |\psi_f\rangle\langle\psi_f| = \hat{1} \qquad ; \text{ completeness}$$

Solution to Problem 10.1 (a) The eigenstates of the hermitian operator $\hat{F}$ are orthonormal

$$\langle\psi_{f'}|\psi_f\rangle = \delta_{f,f'}$$

Hence, the expansion coefficients can be determined to be

$$c_f = \langle\psi_f|\Psi\rangle$$

---

[1] The corresponding statement for the wave functions is obtained by taking the inner product with $|x\rangle$.

(b) Substitute this back in the expansion of the state vector to obtain

$$|\Psi\rangle = \sum_f |\psi_f\rangle\langle\psi_f|\Psi\rangle$$

(c) We conclude that the statement of completeness for the eigenstates of the hermitian operator $\hat{F}$ in abstract Hilbert space can be written as

$$\sum_f |\psi_f\rangle\langle\psi_f| = \hat{1} \qquad ; \text{ completeness}$$

Problem 10.2 (a) If one performs a pure pass measurement at a time t_0 that lets the eigenvalue f through, show that the rescaled reduced wave function at t_0 is[2]

$$\Psi(x, t_0) = \frac{c_f(t_0)}{|c_f(t_0)|}\psi_f(x) \qquad ; t = t_0$$

(b) Give an argument that the reduced wave function at subsequent times is then

$$\Psi(x, t) = \frac{c_f(t_0)}{|c_f(t_0)|}\psi_f(x)\, e^{-iE_f(t-t_0)/\hbar} \qquad ; t \geq t_0$$

Solution to Problem 10.2 (a) The eigenstates of the hermitian operator F satisfy[3]

$$F\psi_f(x) = f\psi_f(x) \qquad ; \text{ eigenfunctions}$$

They are assumed to form a complete set in which the wave function $\Psi(x, t)$ can be expanded

$$\Psi(x, t) = \sum_f c_f(t)\psi_f(x) \qquad ; \text{ complete set}$$

A pure pass measurement of the eigenvalue f at the time t_0 says that the system is now in the state $\psi_f(x)$ with unit probability. For a pure pass measurement, the coefficient $c_f(t_0)$ is unchanged, but it must be *rescaled* to reflect the unit probability. Thus at the time t_0, the measurement has reduced the wave function to

$$\Psi(x, t_0) = \frac{c_f(t_0)}{|c_f(t_0)|}\psi_f(x) \qquad ; t = t_0$$

---

[2]Recall Eq. (8.11).

[3]Here we consider operators F with an infinite range of eigenvalues f, together with a continuous coordinate in the wave function, such as our pairs (p, x), (H, x), and (L_z, ϕ).

(b) If, after the pure pass measurement is made, the system just propagates freely in the state $\Psi_f(x)$, then the wave function is

$$\Psi(x, t) = \Psi(x, t_0)e^{-iE_f(t-t_0)/\hbar}$$

Hence

$$\Psi(x, t) = \frac{c_f(t_0)}{|c_f(t_0)|}\psi_f(x)\,e^{-iE_f(t-t_0)/\hbar} \qquad ;\ t \geq t_0.$$

Chapter 11

Relativity

Problem 11.1 (a) Expand the square-root in Eq. (11.6) to first order, and show that, apart from a constant term m_0c^2 in the energy, one obtains the non-relativistic Schrödinger equation;

(b) What is the first relativistic correction to this Schrödinger equation?

Solution to Problem 11.1

(a) Our first attempt in Eq. (11.6) to incorporate relativity into the Schrödinger equation reads

$$i\hbar\frac{\partial\phi(\vec{x},t)}{\partial t} = \left[-(\hbar c)^2\vec{\nabla}^2 + (m_0c^2)^2\right]^{1/2}\phi(\vec{x},t)$$

The square root can be expanded as

$$\left[-(\hbar c)^2\vec{\nabla}^2 + (m_0c^2)^2\right]^{1/2} = m_0c^2\left[1 - \frac{\hbar^2\vec{\nabla}^2}{(m_0c)^2}\right]^{1/2}$$

$$= m_0c^2 - \frac{\hbar^2\vec{\nabla}^2}{2m_0} - \frac{m_0c^2}{8}\left(\frac{\hbar\vec{\nabla}}{m_0c}\right)^4 + \cdots$$

A constant term in the energy is irrelevant in the non-relativistic theory, and hence the first two terms on the r.h.s. lead to the non-relativistic Schrödinger equation.

(b) The third term on the r.h.s. provides the first relativistic correction to this non-relativistic Schrödinger equation

$$H' = -\frac{m_0c^2}{8}\left(\frac{\hbar\vec{\nabla}}{m_0c}\right)^4$$

Problem 11.2 As in Appendix A, substitute the normal-mode expansion of the scalar field in Eq. (11.23) into the hamiltonian density in Eq. (11.19), do the spatial integrals, and derive the uncoupled oscillator expansion of the energy in Eq. (11.24).

Solution to Problem 11.2 The hamiltonian density for the massive neutral scalar field is given in Eq. (11.19) as

$$
\mathcal{H}(\vec{x}, t) = \frac{1}{2}\left[\frac{\partial \phi(\vec{x}, t)}{\partial t}\right]^2 + \frac{c^2}{2}\left[\vec{\nabla}\phi(\vec{x}, t)\right]^2 + \frac{1}{2}m^2 c^2 \phi^2(\vec{x}, t)
$$

The normal-mode expansion of that field is given in Eq. (11.23)

$$
\phi(\vec{x}, t) = \sum_{\vec{k}} \left(\frac{\hbar}{2\omega_k \Omega}\right)^{1/2}\left[c_{\vec{k}}\, e^{i(\vec{k}\cdot\vec{x}-\omega_k t)} + c_{\vec{k}}^\star\, e^{-i(\vec{k}\cdot\vec{x}-\omega_k t)}\right]
$$

It follows that

$$
\frac{\partial \phi(\vec{x}, t)}{\partial t} = \frac{1}{i}\sum_{\vec{k}} \left(\frac{\hbar\omega_k}{2\Omega}\right)^{1/2}\left[c_{\vec{k}}\, e^{i(\vec{k}\cdot\vec{x}-\omega_k t)} - c_{\vec{k}}^\star\, e^{-i(\vec{k}\cdot\vec{x}-\omega_k t)}\right]
$$

$$
\vec{\nabla}\phi(\vec{x}, t) = i\sum_{\vec{k}} \left(\frac{\hbar}{2\omega_k \Omega}\right)^{1/2}\vec{k}\left[c_{\vec{k}}\, e^{i(\vec{k}\cdot\vec{x}-\omega_k t)} - c_{\vec{k}}^\star\, e^{-i(\vec{k}\cdot\vec{x}-\omega_k t)}\right]
$$

We substitute these expansions in the above and do the integral over all space using the orthonormality of the plane waves in a big box of volume $\Omega = L^3$ with periodic boundary conditions

$$
\frac{1}{\Omega}\int d^3x\, e^{i(\vec{k}-\vec{k}')\cdot\vec{x}} = \delta_{\vec{k},\vec{k}'}
$$

There will be three contributions:
First, one has

$$
\int d^3x\, \frac{1}{2}\left[\frac{\partial \phi(\vec{x}, t)}{\partial t}\right]^2 = \frac{1}{4}\sum_{\vec{k}} \hbar\omega_k \left\{\left(c_{\vec{k}}c_{\vec{k}}^\star + c_{\vec{k}}^\star c_{\vec{k}}\right)\right.
$$
$$
\left. - \left(c_{\vec{k}}c_{-\vec{k}}\, e^{-2i\omega_k t} + c_{\vec{k}}^\star c_{-\vec{k}}^\star\, e^{2i\omega_k t}\right)\right\}
$$

then

$$
\int d^3x\, \frac{c^2}{2}\left[\vec{\nabla}\phi(\vec{x}, t)\right]^2 = \frac{c^2}{4}\sum_{\vec{k}} \frac{\hbar k^2}{\omega_k}\left\{\left(c_{\vec{k}}c_{\vec{k}}^\star + c_{\vec{k}}^\star c_{\vec{k}}\right)\right.
$$
$$
\left. + \left(c_{\vec{k}}c_{-\vec{k}}\, e^{-2i\omega_k t} + c_{\vec{k}}^\star c_{-\vec{k}}^\star\, e^{2i\omega_k t}\right)\right\}
$$

and finally

$$\int d^3x \, \frac{1}{2} m^2 c^2 \phi^2(\vec{x},t) = \frac{m^2 c^2}{4} \sum_{\vec{k}} \frac{\hbar}{\omega_k} \left\{ \left(c_{\vec{k}}^\star c_{\vec{k}}^\star + c_{\vec{k}}^\star c_{\vec{k}} \right) \right.$$
$$\left. + \left(c_{\vec{k}} c_{-\vec{k}} \, e^{-2i\omega_k t} + c_{\vec{k}}^\star c_{-\vec{k}}^\star \, e^{2i\omega_k t} \right) \right\}$$

Now combine these three expressions using the relation for the angular frequency

$$c^2 \left(k^2 + m^2 \right) = \omega_k^2$$

The result is

$$H = \frac{1}{2} \sum_{\vec{k}} \hbar \omega_k \left(c_{\vec{k}}^\star c_{\vec{k}} + c_{\vec{k}} c_{\vec{k}}^\star \right)$$

As was the case for the electromagnetic field, this is just the total energy of the neutral scalar meson system in normal modes

$$H = E$$

While the individual terms oscillate with time, the total energy is a constant of the motion.

Problem 11.3 (a) The canonical momentum density for the neutral massive scalar field is given by

$$\Pi(\vec{x},t) = \frac{\partial \mathcal{L}}{\partial(\partial\phi/\partial t)} = \frac{\partial\phi(\vec{x},t)}{\partial t}$$

The expansion of the scalar field in normal modes is

$$\phi(\vec{x},t) = \sum_{\vec{k}} \left(\frac{\hbar}{2\omega_k \Omega} \right)^{1/2} \left[c_{\vec{k}} \, e^{i(\vec{k}\cdot\vec{x}-\omega_k t)} + c_{\vec{k}}^\dagger \, e^{-i(\vec{k}\cdot\vec{x}-\omega_k t)} \right]$$

Hence, show that the canonical momentum density is

$$\Pi(\vec{x},t) = \frac{1}{i} \sum_{\vec{k}} \left(\frac{\hbar \omega_k}{2\Omega} \right)^{1/2} \left[c_{\vec{k}} \, e^{i(\vec{k}\cdot\vec{x}-\omega_k t)} - c_{\vec{k}}^\dagger \, e^{-i(\vec{k}\cdot\vec{x}-\omega_k t)} \right]$$

(b) Use the commutation relations of the creation and destruction operators to show that the equal-time commutation relation of the field and canonical momentum density is

$$[\Phi(\vec{x},t),\,\Pi(\vec{x}',t)] = \frac{i\hbar}{2}\sum_{\vec{k}}\frac{1}{\Omega}\left[e^{i\vec{k}\cdot(\vec{x}-\vec{x}')} + e^{-i\vec{k}\cdot(\vec{x}-\vec{x}')}\right]$$

(c) Each sum provides an integral representation of the three-dimensional Dirac delta function. Hence show that the canonical commutation relation in continuum mechanics between the field and canonical momentum density is

$$[\Phi(\vec{x},t),\,\Pi(\vec{x}',t)] = i\hbar\,\delta^{(3)}(\vec{x}-\vec{x}')$$

Solution to Problem 11.3 (a) As stated, the canonical momentum density for the neutral massive scalar field is given by

$$\Pi(\vec{x},t) = \frac{\partial\mathcal{L}}{\partial(\partial\phi/\partial t)} = \frac{\partial\phi(\vec{x},t)}{\partial t}$$

With the use of the expansion of the scalar field in normal modes

$$\phi(\vec{x},t) = \sum_{\vec{k}}\left(\frac{\hbar}{2\omega_k\Omega}\right)^{1/2}\left[c_{\vec{k}}\,e^{i(\vec{k}\cdot\vec{x}-\omega_k t)} + c_{\vec{k}}^{\dagger}\,e^{-i(\vec{k}\cdot\vec{x}-\omega_k t)}\right]$$

this gives the canonical momentum density as

$$\Pi(\vec{x},t) = \frac{1}{i}\sum_{\vec{k}}\left(\frac{\hbar\omega_k}{2\Omega}\right)^{1/2}\left[c_{\vec{k}}\,e^{i(\vec{k}\cdot\vec{x}-\omega_k t)} - c_{\vec{k}}^{\dagger}\,e^{-i(\vec{k}\cdot\vec{x}-\omega_k t)}\right]$$

(b) The use of the commutation relations for the creation and destruction operators

$$[c_{\vec{k}},\,c_{\vec{k}'}^{\dagger}] = \delta_{\vec{k},\vec{k}'}$$
$$[c_{\vec{k}}^{\dagger},\,c_{\vec{k}'}^{\dagger}] = [c_{\vec{k}},\,c_{\vec{k}'}] = 0$$

then gives

$$[\Phi(\vec{x},t),\,\Pi(\vec{x}',t)] = \frac{i\hbar}{2}\sum_{\vec{k}}\frac{1}{\Omega}\left[e^{i\vec{k}\cdot(\vec{x}-\vec{x}')} + e^{-i\vec{k}\cdot(\vec{x}-\vec{x}')}\right]$$

(c) Each term in this relation is just the three-dimensional generalization of the completeness relation in Prob. 9.4

$$\frac{1}{L}\sum_{k}e^{ik(x-x')} = \frac{1}{L}\sum_{k}e^{-ik(x-x')} = \delta(x-x')$$

Hence

$$[\Phi(\vec{x},t), \Pi(\vec{x}',t)] = i\hbar\, \delta^{(3)}(\vec{x}-\vec{x}')$$

In reverse, this is the canonical equal-time commutation relation between the field and canonical momentum density applied in continuum mechanics to produce quantum field theory.

Problem 11.4 Suppose we include the additional nonlinear self-coupling of the neutral, massive scalar meson field of Eq. (11.28) in the lagrangian density

$$\mathcal{L}_1(\phi) = -\frac{\lambda}{4!}\phi^4$$

(a) Show that since there are no additional derivative terms, the canonical momentum density, and canonical quantization procedure follow exactly as in the previous Prob. 11.3;

(b) Show the additional term in the hamiltonian density is simply

$$\mathcal{H}_1(\phi) = \frac{\lambda}{4!}\phi^4$$

(c) Substitute the normal-mode expansion of Eq. (12.161), and enumerate the processes described by this hamiltonian density;

(d) What picture are we in?

Solution to Problem 11.4 The full lagrangian density for the neutral, massive scalar field with this self-coupling follows from Eq. (11.15) as

$$\mathcal{L} = -\frac{c^2}{2}\left(\frac{\partial\phi}{\partial x_\mu}\right)^2 - \frac{1}{2}m^2c^2\phi^2 - \frac{\lambda}{4!}\phi^4$$

(a) Since there are no new derivative terms, the canonical momentum density is still given by Eq. (11.17)

$$\Pi(\vec{x},t) = \frac{\partial\mathcal{L}}{\partial(\partial\phi/\partial t)} = \frac{\partial\phi(\vec{x},t)}{\partial t}$$

The canonical quantization of the field then follows exactly as in Prob. 11.3, and the field and momentum density are expressed in terms of the creation and destruction operators as

$$\phi(\vec{x}, t) = \sum_{\vec{k}} \left(\frac{\hbar}{2\omega_k \Omega} \right)^{1/2} \left[c_{\vec{k}} \, e^{i(\vec{k}\cdot\vec{x} - \omega_k t)} + c_{\vec{k}}^\dagger \, e^{-i(\vec{k}\cdot\vec{x} - \omega_k t)} \right]$$

$$\Pi(\vec{x}, t) = \frac{1}{i} \sum_{\vec{k}} \left(\frac{\hbar \omega_k}{2\Omega} \right)^{1/2} \left[c_{\vec{k}} \, e^{i(\vec{k}\cdot\vec{x} - \omega_k t)} - c_{\vec{k}}^\dagger \, e^{-i(\vec{k}\cdot\vec{x} - \omega_k t)} \right]$$

This is what we obtained just working with the hamiltonian $\hat{H}_0$.

(b) In passing from the lagrangian density to the hamiltonian density, since there are no new derivatives involved, one has simply

$$\mathcal{H}_1(\phi) = -\mathcal{L}_1(\phi) = \frac{\lambda}{4!} \phi^4$$

(c) Substitution of the field expansion into the interaction gives

$$\hat{\mathcal{H}}_1(\phi) = \frac{\lambda}{4!} \left\{ \sum_{\vec{k}} \left(\frac{\hbar}{2\omega_k \Omega} \right)^{1/2} \left[c_{\vec{k}} \, e^{i(\vec{k}\cdot\vec{x} - \omega_k t)} + c_{\vec{k}}^\dagger \, e^{-i(\vec{k}\cdot\vec{x} - \omega_k t)} \right] \right\}^4$$

Let us characterize the types of terms in this interaction (see figure below):

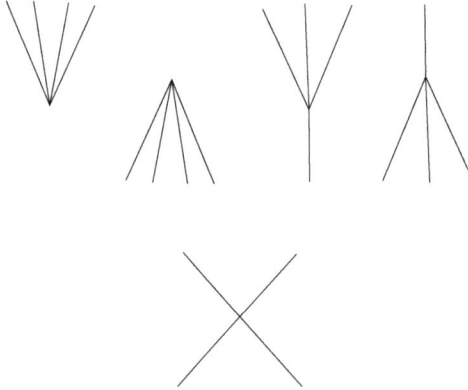

Fig. 11.1   Types of terms in the interaction $(\lambda/4!)\phi^4$.

- There is a term in $c^\dagger c^\dagger c^\dagger c^\dagger$. This creates four mesons from the vacuum and only contributes to virtual processes;
- There is a similar term in $cccc$;

- There are terms in $c^\dagger c^\dagger c^\dagger c$. These contribute to the self-energy of the scalar meson in higher order;
- There are similar terms in $c^\dagger ccc$;
- Finally, there are terms in $c^\dagger c^\dagger cc$ that contribute to the real *scattering* of the scalar mesons.

(d) Since all the creation and destruction operators here carry the free-field time-dependence, we are working in the *interaction picture*.

Problem 11.5 Let $\underline{m}$ denote an $n \times n$ matrix

$$\underline{m} = \begin{bmatrix} m_{11} & m_{12} & \cdots & m_{1n} \\ m_{21} & m_{22} & \cdots & m_{2n} \\ \vdots & \vdots & \vdots & \vdots \\ m_{n1} & m_{n2} & \cdots & m_{nn} \end{bmatrix}$$

Label an element of $\underline{m}$ by m_{jk} where j indicates the row and k indicates the column.[1] Introduce the convention that repeated Latin indices are summed from 1 to n. The element of the matrix product of two such matrices $\underline{a}$ and $\underline{b}$ is then given by

$$[\underline{a}\,\underline{b}]_{jk} = a_{jl}b_{lk}$$

Pick the row for a_{jl} and the column for b_{lk}; then multiply the elements together and add them up.

(a) Evaluate the following matrix products

$$\begin{bmatrix} 0 & 1 \\ 1 & 0 \end{bmatrix}\begin{bmatrix} 0 & -i \\ i & 0 \end{bmatrix} \quad ; \quad \begin{bmatrix} 0 & 1 \\ 1 & 0 \end{bmatrix}\begin{bmatrix} 1 & 0 \\ 0 & -1 \end{bmatrix} \quad ; \quad \begin{bmatrix} 0 & -i \\ i & 0 \end{bmatrix}\begin{bmatrix} 1 & 0 \\ 0 & -1 \end{bmatrix}$$

(b) Let $\underline{\psi}$ denote an n-component column vector

$$\underline{\psi} = \begin{bmatrix} \psi_1 \\ \vdots \\ \psi_n \end{bmatrix}$$

The matrix product $\underline{m}\,\underline{\psi}$ is given by

$$[\underline{m}\,\underline{\psi}]_j = m_{jk}\psi_k$$

---

[1] I cannot believe, with all the checking, that this statement got transposed in the text; my sincere apologies.

Evaluate the following matrix products[2]

$$\begin{bmatrix} 0 & 1 \\ 1 & 0 \end{bmatrix} \begin{bmatrix} 1 \\ 0 \end{bmatrix} \quad ; \quad \begin{bmatrix} 0 & 1 \\ 1 & 0 \end{bmatrix} \begin{bmatrix} 0 \\ 1 \end{bmatrix} \quad ; \quad \begin{bmatrix} 0 & -i \\ i & 0 \end{bmatrix} \begin{bmatrix} 1 \\ 0 \end{bmatrix}$$

Solution to Problem 11.5 Remember — to compute the (j,k)th element of the matrix product, go to the jth row of the first matrix and the kth column of the second. Work down the line for both, multiplying the elements together and summing them.

(a) It follows that

$$\begin{bmatrix} 0 & 1 \\ 1 & 0 \end{bmatrix} \begin{bmatrix} 0 & -i \\ i & 0 \end{bmatrix} = \begin{bmatrix} i & 0 \\ 0 & -i \end{bmatrix}$$

$$\begin{bmatrix} 0 & 1 \\ 1 & 0 \end{bmatrix} \begin{bmatrix} 1 & 0 \\ 0 & -1 \end{bmatrix} = \begin{bmatrix} 0 & -1 \\ 1 & 0 \end{bmatrix}$$

$$\begin{bmatrix} 0 & -i \\ i & 0 \end{bmatrix} \begin{bmatrix} 1 & 0 \\ 0 & -1 \end{bmatrix} = \begin{bmatrix} 0 & i \\ i & 0 \end{bmatrix}$$

(b) In a similar fashion, one has

$$\begin{bmatrix} 0 & 1 \\ 1 & 0 \end{bmatrix} \begin{bmatrix} 1 \\ 0 \end{bmatrix} = \begin{bmatrix} 0 \\ 1 \end{bmatrix}$$

$$\begin{bmatrix} 0 & 1 \\ 1 & 0 \end{bmatrix} \begin{bmatrix} 0 \\ 1 \end{bmatrix} = \begin{bmatrix} 1 \\ 0 \end{bmatrix}$$

$$\begin{bmatrix} 0 & -i \\ i & 0 \end{bmatrix} \begin{bmatrix} 1 \\ 0 \end{bmatrix} = \begin{bmatrix} 0 \\ i \end{bmatrix}$$

Problem 11.6 The Pauli matrices are defined in Eqs. (11.54).

(a) Show the matrix product of Pauli matrices satisfies

$$\sigma_i \sigma_j = i\varepsilon_{ijk}\sigma_k \qquad ; \ i \neq j$$

where ε_{ijk} is the completely antisymmetric Levi-Civita tensor;[3]

(b) Show the Pauli matrices satisfy the following anticommutation relation

$$\sigma_i \sigma_j + \sigma_j \sigma_i = 2\delta_{ij}$$

---

[2]The center one has been modified.
[3]The *factor* of i here is $i = \sqrt{-1}$.

Solution to Problem 11.6 As defined in Eqs. (11.54), the 2×2 Pauli matrices are given by

$$\sigma_x = \begin{pmatrix} 0 & 1 \\ 1 & 0 \end{pmatrix} \quad ; \sigma_y = \begin{pmatrix} 0 & -i \\ i & 0 \end{pmatrix} \quad ; \sigma_z = \begin{pmatrix} 1 & 0 \\ 0 & -1 \end{pmatrix} \quad ; \text{ Pauli matrices}$$

With the rules for matrix multiplication discussed in the previous problem, we can work out several of the matrix products. For example[4]

$$\sigma_x \sigma_x = \begin{pmatrix} 0 & 1 \\ 1 & 0 \end{pmatrix} \begin{pmatrix} 0 & 1 \\ 1 & 0 \end{pmatrix} = \begin{pmatrix} 1 & 0 \\ 0 & 1 \end{pmatrix} = 1$$

$$\sigma_x \sigma_y = \begin{pmatrix} 0 & 1 \\ 1 & 0 \end{pmatrix} \begin{pmatrix} 0 & -i \\ i & 0 \end{pmatrix} = \begin{pmatrix} i & 0 \\ 0 & -i \end{pmatrix} = i\sigma_z$$

$$\sigma_y \sigma_x = \begin{pmatrix} 0 & -i \\ i & 0 \end{pmatrix} \begin{pmatrix} 0 & 1 \\ 1 & 0 \end{pmatrix} = \begin{pmatrix} -i & 0 \\ 0 & i \end{pmatrix} = -i\sigma_z = -\sigma_x \sigma_y$$

Similarly

$$\sigma_z \sigma_z = \begin{pmatrix} 1 & 0 \\ 0 & -1 \end{pmatrix} \begin{pmatrix} 1 & 0 \\ 0 & -1 \end{pmatrix} = \begin{pmatrix} 1 & 0 \\ 0 & 1 \end{pmatrix} = 1$$

$$\sigma_z \sigma_x = \begin{pmatrix} 1 & 0 \\ 0 & -1 \end{pmatrix} \begin{pmatrix} 0 & 1 \\ 1 & 0 \end{pmatrix} = \begin{pmatrix} 0 & 1 \\ -1 & 0 \end{pmatrix} = i\sigma_y = -\sigma_x \sigma_z$$

And finally

$$\sigma_y \sigma_y = \begin{pmatrix} 0 & -i \\ i & 0 \end{pmatrix} \begin{pmatrix} 0 & -i \\ i & 0 \end{pmatrix} = \begin{pmatrix} 1 & 0 \\ 0 & 1 \end{pmatrix} = 1$$

$$\sigma_y \sigma_z = \begin{pmatrix} 0 & -i \\ i & 0 \end{pmatrix} \begin{pmatrix} 1 & 0 \\ 0 & -1 \end{pmatrix} = \begin{pmatrix} 0 & i \\ i & 0 \end{pmatrix} = i\sigma_x = -\sigma_z \sigma_y$$

(a) It follows that the matrix product of Pauli matrices satisfies[5]

$$\sigma_i \sigma_j = i\varepsilon_{ijk}\sigma_k \qquad ; i \neq j$$

where ε_{ijk} is the completely antisymmetric Levi-Civita tensor.

(b) The Pauli matrices similarly satisfy the following anticommutation relation

$$\sigma_i \sigma_j + \sigma_j \sigma_i = 2\delta_{ij}$$

---

[4] We suppress the unit matrix in the results.
[5] Note that in this discussion, the indices $(1, 2, 3)$ are the same as (x, y, z).

Problem 11.7 The *standard representation* of the Dirac matrices in 2×2 form is

$$\vec{\alpha} = \begin{pmatrix} 0 & \vec{\sigma} \\ \vec{\sigma} & 0 \end{pmatrix} \quad ; \; \beta = \begin{pmatrix} 1 & 0 \\ 0 & -1 \end{pmatrix}$$

where $\vec{\sigma} = (\sigma_x, \sigma_y, \sigma_z)$ are the Pauli matrices defined in Eqs. (11.54).

(a) Show the rules for matrix multiplication of the 4×4 matrices are satisfied by using matrix multiplication rules for the 2×2 submatrices;

(b) Show the following relations are satisfied by the Dirac matrices[6]

$$\beta \alpha^k + \alpha^k \beta = 0$$
$$\alpha^k \alpha^l + \alpha^l \alpha^k = 2\delta_{kl}$$
$$\beta^2 = 1$$

Solution to Problem 11.7 (a) Suppose one has a 4×4 matrix written in 2×2 form as

$$\underline{M} = \begin{pmatrix} \underline{a} & \underline{b} \\ \underline{c} & \underline{d} \end{pmatrix}$$

The matrix product of two such matrices in 2×2 form is

$$\underline{M}^1 \underline{M}^2 = \begin{pmatrix} \underline{a}^1 & \underline{b}^1 \\ \underline{c}^1 & \underline{d}^1 \end{pmatrix} \begin{pmatrix} \underline{a}^2 & \underline{b}^2 \\ \underline{c}^2 & \underline{d}^2 \end{pmatrix} = \begin{pmatrix} \underline{a}^1\underline{a}^2 + \underline{b}^1\underline{c}^2 & \underline{a}^1\underline{b}^2 + \underline{b}^1\underline{d}^2 \\ \underline{c}^1\underline{a}^2 + \underline{d}^1\underline{c}^2 & \underline{c}^1\underline{b}^2 + \underline{d}^1\underline{d}^2 \end{pmatrix}$$

Now consider a given element of this 4×4 product, say $(2, 3)$

$$\left[\underline{M}^1 \underline{M}^2\right]_{23} = \left[\underline{M}^1\right]_{2i} \left[\underline{M}^2\right]_{i3} =$$
$$\left[\underline{M}^1\right]_{21} \left[\underline{M}^2\right]_{13} + \left[\underline{M}^1\right]_{22} \left[\underline{M}^2\right]_{23} + \left[\underline{M}^1\right]_{23} \left[\underline{M}^2\right]_{33} + \left[\underline{M}^1\right]_{24} \left[\underline{M}^2\right]_{43}$$

From the definition of $\underline{M}$, this is

$$\left[\underline{M}^1 \underline{M}^2\right]_{23} = a^1_{21} b^2_{11} + a^1_{22} b^2_{21} + b^1_{21} d^2_{11} + b^1_{22} d^2_{21}$$

For 2×2 matrices, this is

$$a^1_{21} b^2_{11} + a^1_{22} b^2_{21} + b^1_{21} d^2_{11} + b^1_{22} d^2_{21} = \left[\underline{a}^1\underline{b}^2 + \underline{b}^1\underline{d}^2\right]_{21}$$

This is the correct element $\left[\underline{M}^1 \underline{M}^2\right]_{23}$ of the matrix product $\left[\underline{M}^1 \underline{M}^2\right]$ as written in 2×2 form.

---

[6]Once again, we do not underline the Pauli or Dirac matrices, and there is a suppressed unit matrix on the r.h.s. of Eqs. (12.174) and (12.176); also, $\alpha^k \equiv \alpha_k$.

(b) We first confirm that the Dirac matrices are hermitian, that is, that each is equal to its complex conjugate transpose. The Pauli matrices are evidently hermitian

$$\vec{\sigma}^\dagger = \vec{\sigma}$$

It follows that the Dirac matrices are then also hermitian

$$\vec{\alpha}^\dagger = \begin{pmatrix} 0 & \vec{\sigma}^\dagger \\ \vec{\sigma}^\dagger & 0 \end{pmatrix} = \begin{pmatrix} 0 & \vec{\sigma} \\ \vec{\sigma} & 0 \end{pmatrix} = \vec{\alpha} \qquad\qquad ; \beta^\dagger = \begin{pmatrix} 1 & 0 \\ 0 & -1 \end{pmatrix} = \beta$$

Now let us look at the anticommutation relations for the Dirac matrices in 2×2 form

$$\beta\vec{\alpha} + \vec{\alpha}\beta = \begin{pmatrix} 1 & 0 \\ 0 & -1 \end{pmatrix} \begin{pmatrix} 0 & \vec{\sigma} \\ \vec{\sigma} & 0 \end{pmatrix} + \begin{pmatrix} 0 & \vec{\sigma} \\ \vec{\sigma} & 0 \end{pmatrix} \begin{pmatrix} 1 & 0 \\ 0 & -1 \end{pmatrix}$$

$$= \begin{pmatrix} 0 & \vec{\sigma} \\ -\vec{\sigma} & 0 \end{pmatrix} + \begin{pmatrix} 0 & -\vec{\sigma} \\ \vec{\sigma} & 0 \end{pmatrix} = 0$$

Furthermore

$$\alpha^k\alpha^l + \alpha^l\alpha^k = \begin{pmatrix} 0 & \sigma_k \\ \sigma_k & 0 \end{pmatrix} \begin{pmatrix} 0 & \sigma_l \\ \sigma_l & 0 \end{pmatrix} + \begin{pmatrix} 0 & \sigma_l \\ \sigma_l & 0 \end{pmatrix} \begin{pmatrix} 0 & \sigma_k \\ \sigma_k & 0 \end{pmatrix}$$

$$= \begin{pmatrix} \sigma_k\sigma_l + \sigma_l\sigma_k & 0 \\ 0 & \sigma_k\sigma_l + \sigma_l\sigma_k \end{pmatrix} = 2\delta_{kl} \begin{pmatrix} 1 & 0 \\ 0 & 1 \end{pmatrix}$$

In addition

$$\beta^2 = \begin{pmatrix} 1 & 0 \\ 0 & -1 \end{pmatrix} \begin{pmatrix} 1 & 0 \\ 0 & -1 \end{pmatrix} = \begin{pmatrix} 1 & 0 \\ 0 & 1 \end{pmatrix}$$

Thus, in *summary*, with our notation, the anticommutation relations for the Dirac matrices are

$$\beta\alpha^k + \alpha^k\beta = 0$$
$$\alpha^k\alpha^l + \alpha^l\alpha^k = 2\delta_{kl}$$
$$\beta^2 = 1$$

Problem 11.8 Use the analysis in Sec. 11.3.1 to show that the non-relativistic limit of the positive-energy Dirac spinor for a free particle takes the form

$$u_\lambda(\vec{k}) = \begin{bmatrix} \phi_\lambda \\ (\vec{\sigma}\cdot\vec{p}/2m_0c)\phi_\lambda \end{bmatrix} \qquad ; \ \phi_\uparrow = \begin{pmatrix} 1 \\ 0 \end{pmatrix}$$

$$; \ \phi_\downarrow = \begin{pmatrix} 0 \\ 1 \end{pmatrix}$$

Solution to Problem 11.8 In the non-relativistic reduction of the Dirac equation examined in Sec. 11.3.1, it is shown in Eq. (11.66) that

$$\frac{\vec{p}^{\,2}}{2m_0}\phi = -\frac{\hbar^2\nabla^2}{2m_0}\phi = \varepsilon\phi \qquad ; \ \text{Schrödinger equation}$$

Here ϕ is a two-component wave function describing the spin. We have shown in Prob. 8.1 that the two wave functions describing spin-up and spin-down along the z-axis are given by

$$\phi_\uparrow = \begin{pmatrix} 1 \\ 0 \end{pmatrix} \qquad ; \ \phi_\downarrow = \begin{pmatrix} 0 \\ 1 \end{pmatrix}$$

It is also shown in Sec. 11.3.1 that in the non-relativistic limit, the bottom two components of the Dirac spinor are given by

$$\chi = \frac{\vec{\sigma}\cdot\vec{p}}{2m_0c}\phi$$

It follows that in the non-relativistic limit, the Dirac spinors for a spin-1/2 particle take the form[7]

$$u_\lambda(\vec{k}) = \begin{bmatrix} \phi_\lambda \\ (\vec{\sigma}\cdot\vec{p}/2m_0c)\phi_\lambda \end{bmatrix} \qquad ; \ \phi_\uparrow = \begin{pmatrix} 1 \\ 0 \end{pmatrix}$$

$$; \ \phi_\downarrow = \begin{pmatrix} 0 \\ 1 \end{pmatrix}$$

where $\lambda = (\uparrow, \downarrow)$ represents the projection of the spin along the z-axis.

Problem 11.9 If the vector potential $\vec{A}$ in Eq. (11.69) is included in the non-relativistic reduction in Eq. (11.66), one has

$$\frac{1}{2m_0}\vec{\sigma}\cdot(\vec{p}-e\vec{A})\,\vec{\sigma}\cdot(\vec{p}-e\vec{A})\,\phi = \varepsilon\phi$$

---

[7]In the plane-wave state $\vec{p} = \hbar\vec{k}$; more generally, $\vec{p} = (\hbar/i)\vec{\nabla}$.

With our convention that repeated Latin indices are summed from 1 to 3, and the results in Prob. 11.6, one has

$$\vec{\sigma} \cdot (\vec{p} - e\vec{A})\, \vec{\sigma} \cdot (\vec{p} - e\vec{A}) = \sigma_i \sigma_j (p - eA)_i (p - eA)_j$$
$$= (\delta_{ij} + i\varepsilon_{ijk}\sigma_k)(p - eA)_i (p - eA)_j$$

(a) Show

$$i\sigma_k \varepsilon_{kij}(p - eA)_i (p - eA)_j = -e\hbar\, \sigma_k \varepsilon_{kij}(\nabla_i A_j)$$
$$= -e\hbar\, \vec{\sigma} \cdot (\vec{\nabla} \times \vec{A})$$
$$= -e\hbar\, \vec{\sigma} \cdot \vec{B}$$

(b) Hence, conclude that the above non-relativistic reduction takes the form

$$\left[\frac{1}{2m_0}(\vec{p} - e\vec{A})^2 - \frac{e\hbar}{2m_0}\vec{\sigma} \cdot \vec{B}\right]\phi = \varepsilon\phi$$

(c) From this, conclude that a Dirac particle has the magnetic moment in Eq. (11.70).

Solution to Problem 11.9 The non-relativistic reduction of the Dirac equation is discussed in Sec. 11.3.1, and the coupled equations for the upper and lower two-component pairs are obtained in Eqs. (11.59).

If the vector potential $\vec{A}$ in Eq. (11.69) is included in the non-relativistic reduction in Eq. (11.66), one has

$$\frac{1}{2m_0}\vec{\sigma} \cdot (\vec{p} - e\vec{A})\, \vec{\sigma} \cdot (\vec{p} - e\vec{A})\, \phi = \varepsilon\phi$$

where ϕ is the two-component spinor of the previous problem.

(a) With our convention that repeated Latin indices are summed from 1 to 3, one has

$$\vec{\sigma} \cdot (\vec{p} - e\vec{A})\, \vec{\sigma} \cdot (\vec{p} - e\vec{A}) = \sigma_i \sigma_j (p - eA)_i (p - eA)_j$$

The results in Prob. 11.6 can be combined to give[8]

$$\sigma_i \sigma_j \equiv \frac{1}{2}(\sigma_i \sigma_j + \sigma_j \sigma_i) + \frac{1}{2}(\sigma_i \sigma_j - \sigma_j \sigma_i)$$
$$= \delta_{ij} + i\varepsilon_{ijk}\sigma_k$$

---

[8]We again suppress the unit matrix in the first term.

Hence

$$\vec{\sigma} \cdot (\vec{p} - e\vec{A}) \, \vec{\sigma} \cdot (\vec{p} - e\vec{A}) = (\vec{p} - e\vec{A})^2 + i\varepsilon_{ijk}\sigma_k(p - eA)_i(p - eA)_j$$

Since $p_i = (\hbar/i)\nabla_i$, the last term can be evaluated as

$$\begin{aligned} i\sigma_k\varepsilon_{kij}(p - eA)_i(p - eA)_j &= -e\hbar\,\sigma_k\varepsilon_{kij}(\nabla_i A_j) \\ &= -e\hbar\,\vec{\sigma}\cdot(\vec{\nabla}\times\vec{A}) \\ &= -e\hbar\,\vec{\sigma}\cdot\vec{B} \end{aligned}$$

(b) From this, we conclude that the above non-relativistic reduction of the Dirac equation takes the form

$$\left[\frac{1}{2m_0}(\vec{p} - e\vec{A})^2 - \frac{e\hbar}{2m_0}\vec{\sigma}\cdot\vec{B}\right]\phi = \varepsilon\phi$$

(c) It follows that as a Dirac particle, the electron has the magnetic moment in Eq. (11.70)

$$\vec{\mu}_{\text{el}} = \frac{e\hbar}{2m_e}2\vec{S} \qquad\qquad ; \, \vec{S} = \frac{1}{2}\vec{\sigma}$$

Problem 11.10 (a) Consider the Dirac equation for a particle in a static, central, electric field given by $\vec{E} = -\vec{\nabla}\Phi(r)$. Show that in this case the Dirac equation can be reduced to

$$\left[c\vec{\sigma}\cdot\vec{p}\,\frac{1}{2m_0c^2 + \varepsilon - e\Phi(r)}\,c\vec{\sigma}\cdot\vec{p} + e\Phi(r)\right]\phi = \varepsilon\phi \qquad ; \, \varepsilon \equiv E - m_0c^2$$

(b) Show that[9]

$$\vec{\sigma}\cdot\vec{p}\,\Phi(r)\,\vec{\sigma}\cdot\vec{p} = \Phi\vec{p}^2 + (\vec{p}\Phi)\cdot\vec{p} + \frac{\hbar}{r}\left(\frac{d\Phi}{dr}\right)\vec{\sigma}\cdot(\vec{r}\times\vec{p})$$

(c) Hence conclude that if $|\varepsilon|/2m_0c^2 \ll 1$, and $e\Phi\left\{1 + O[(p/m_0c)^2]\right\} \approx e\Phi$, then the Dirac equation reduces to the following Schrödinger equation for the upper components ϕ

$$H\phi = \varepsilon\phi$$

$$H = \frac{\vec{p}^2}{2m_0} + e\Phi(r) + \frac{e\hbar^2}{(2m_0c)^2}\frac{1}{r}\left(\frac{d\Phi}{dr}\right)\vec{\sigma}\cdot\vec{l}$$

---

[9]Note that $\vec{\nabla}\Phi(r) = (\vec{r}/r)(d\Phi/dr)$.

Here the spin-dependent contribution has been retained, and the angular momentum identified as $\vec{r} \times \vec{p} = \hbar \vec{l}$.

(d) Use this result to obtain Eq. (11.71).

Solution to Problem 11.10 (a) The Dirac eigenfunction equation now reads

$$\left[c\vec{\alpha} \cdot \vec{p} + \beta m_0 c^2 + e\Phi \right] \psi = E\psi$$

The rewriting of this equation in two-component form as in Eq. (11.59) then gives

$$c\vec{\sigma} \cdot \vec{p}\chi + \left[e\Phi(r) + m_0 c^2 \right] \phi = E\phi$$
$$c\vec{\sigma} \cdot \vec{p}\phi + \left[e\Phi(r) - m_0 c^2 \right] \chi = E\chi$$

Solution for χ in the second equation, and substitution in the first, then leads to the following equation for ϕ

$$\left[c\vec{\sigma} \cdot \vec{p} \frac{1}{2m_0 c^2 + \varepsilon - e\Phi(r)} c\vec{\sigma} \cdot \vec{p} + e\Phi(r) \right] \phi = \varepsilon\phi \qquad ; \varepsilon \equiv E - m_0 c^2$$

(b) Consider

$$\vec{\sigma} \cdot \vec{p}\,\Phi(r)\,\vec{\sigma} \cdot \vec{p} = -\hbar^2 \sigma_i \sigma_j \nabla_i \Phi(r) \nabla_j$$
$$= -\hbar^2 \sigma_i \sigma_j \left\{ \Phi(r) \nabla_i \nabla_j + [\nabla_i \Phi(r)] \nabla_j \right\}$$

Use, from the previous problem,

$$\sigma_i \sigma_j = \delta_{ij} + i\varepsilon_{ijk} \sigma_k$$

Hence

$$\vec{\sigma} \cdot \vec{p}\,\Phi(r)\,\vec{\sigma} \cdot \vec{p} = \Phi \vec{p}^2 + (\vec{p}\Phi) \cdot \vec{p} - i\hbar^2 \varepsilon_{ijk} \sigma_k [\nabla_i \Phi(r)] \nabla_j$$

We know that

$$\vec{\nabla}\Phi(r) = \frac{\vec{r}}{r} \frac{d\Phi(r)}{dr}$$

It follows that the last term in the above is

$$-i\hbar^2 \varepsilon_{ijk} \sigma_k [\nabla_i \Phi(r)] \nabla_j = \frac{\hbar}{r} \frac{d\Phi(r)}{dr} \varepsilon_{ijk} \sigma_k r_i p_j$$
$$= \frac{\hbar}{r} \frac{d\Phi(r)}{dr} \vec{\sigma} \cdot (\vec{r} \times \vec{p})$$

Thus, finally,

$$\vec{\sigma} \cdot \vec{p}\, \Phi(r)\, \vec{\sigma} \cdot \vec{p} = \Phi \vec{p}^2 + (\vec{p}\Phi) \cdot \vec{p} + \frac{\hbar}{r}\left(\frac{d\Phi}{dr}\right)\vec{\sigma} \cdot (\vec{r} \times \vec{p})$$

(c) Now expand the denominator of the expression in part (a), and assume that $|\varepsilon|/2m_0c^2 \ll 1$, and $e\Phi\left\{1 + O[(p/m_0c)^2]\right\} \approx e\Phi$. The Dirac equation then reduces to the following Schrödinger equation for the upper components ϕ

$$H\phi = \varepsilon\phi$$
$$H = \frac{\vec{p}^2}{2m_0} + e\Phi(r) + \frac{e\hbar^2}{(2m_0c)^2}\frac{1}{r}\left(\frac{d\Phi}{dr}\right)\vec{\sigma} \cdot \vec{l}$$

Here the spin-dependent contribution has been retained, and the angular momentum identified as $\vec{r} \times \vec{p} = \hbar\vec{l}$.

(d) We conclude that in a static, central, electric field described by $\Phi(r)$, the electron experiences a *spin-orbit interaction*

$$V_{\text{SO}} = e\left(\frac{\hbar}{2m_ec}\right)^2\frac{1}{r}\left(\frac{d\Phi}{dr}\right)2\vec{S} \cdot \vec{l}$$

where the orbital angular momentum is $\hbar\vec{l}$. This is again in accord with the experimental observation.

Appendix A

Electromagnetic Field in Normal Modes

See Problem 11.2.

Bibliography

Abers, E. S., and Lee, B. W., (1973). *Phys. Rep.* **9**, 1

Amore, P., and Walecka, J. D., (2013). *Introduction to Modern Physics: Solutions to Problems*, World Scientific Publishing Company, Singapore

Amore, P., and Walecka, J. D., (2014). *Topics in Modern Physics: Solutions to Problems*, World Scientific Publishing Company, Singapore

Amore, P., and Walecka, J. D., (2015). *Advanced Modern Physics: Solutions to Problems*, World Scientific Publishing Company, Singapore

Bernhardt, C., (2019). *Quantum Computing for Everyone*, M.I.T. Press. Cambridge, MA

Bjorken, J. D., and Drell, S. D., (1964). *Relativistic Quantum Mechanics*, McGraw-Hill, New York, NY

Bjorken, J. D., and Drell, S. D., (1965). *Relativistic Quantum Fields*, McGraw-Hill, New York, NY

Born, M., (1926). *Zeit. f. Physik* **37**, 863; (1927) *Nature*, **119**, 354

Dirac, P. A. M., (1926). *Proc. Roy. Soc.*, **117**, 610

Dirac, P. A. M., (1930). *The Principles of Quantum Mechanics*, Clarendon Press, Oxford, UK

Fetter, A. L., and Walecka, J. D., (2003). *Theoretical Mechanics of Particles and Continua*, McGraw-Hill, New York (1980); reissued by Dover Publications, Mineola, New York

Fetter, A. L., and Walecka, J. D., (2003a). *Quantum Theory of Many-Particle Systems*, McGraw-Hill, New York (1971); reissued by Dover Publications, Mineola, New York

Fetter, A. L., and Walecka, J. D., (2006). *Nonlinear Mechanics: A Supplement to Theoretical Mechanics of Particles and Continua*, Dover Publications, Mineola, New York

Freedman, R., Ruskell, T., Keston, P. M., and Tauck, D. L., (2013). *College Physics*, W. H. Freeman, San Francisco, CA

Feynman, R. P., Hibbs, A. R., and Steyer, D. F., (2010). *Quantum Mechanics and Path Integrals, emended ed.*, Dover Publications, Mineola, NY

Goldstein, H., Poole, C. P., and Safko, J., (2011). *Classical Mechanics, 3rd international economy ed.*, Pearson Education, London, UK

Gottfried, K., and Yan, T.-M., (2004). *Quantum Mechanics: Fundamentals, 2nd ed.*, Springer, New York, NY

Halliday, D., Resnick, R., and Walker, J., (2013). *Fundamentals of Physics, 10th ed.*, J. Wiley and Sons, New York, NY

Itzykson, C., and Zuber, J.-B., (1980). *Quantum Field Theory*, McGraw-Hill, New York, NY

Kibble, T. W. B., and Berkshire, F. H., (2004). *Classical Mechanics, 5th ed.*, Imperial College, London, UK

Kleppner, D., and Kolenkow, R., (2013). *An Introduction to Mechanics, 2nd ed.*, Cambridge University Press, Cambridge, UK

Landau, L. D., and Lifshitz, E. M., (1976). *Mechanics*, Butterworth-Heinemann, Oxford, UK

Landau, L. D., and Lifshitz, E. M., (1981). *Quantum Mechanics, 3rd ed.*, Butterworth-Heinemann, New York, NY

Merzbacher, E., (1997). *Quantum Mechanics*, John Wiley and Sons, New York, NY

Morin, D., (2008). *Introduction to Classical Mechanics: With Problems and Solutions*, Cambridge University Press, Cambridge, UK

Ohanian, H. C., (1985). *Physics*, W. W. Norton and Co., New York, NY

Ohanian, H. C., (1995). *Modern Physics, 2nd ed.*, Prentice-Hall, Upper Saddle River, NJ

Planck, M., (1901). *Annalen der Physik* **4**, 533

Schiff, L. I., (1968). *Quantum Mechanics, 3rd ed.*, McGraw-Hill, New York, NY

Shankar, R., (1994). *Principles of Quantum Mechanics, 2nd ed.*, Springer, New York, NY

Schrödinger, E., (1926). *Annalen der Physik*, **79**, 489; **81**, 109

Taylor, J. R., (2004). *Classical Mechanics*, University Science Books, Sausilito, CA

Thornton, S. T., and Marion, J. B., (2012). *Classical Dynamics of Particles and Systems, 5th ed.*, Cengage Learning, Boston, MA

Walecka, J. D., (2000). *Fundamentals of Statistical Mechanics: Manuscript and Notes of Felix Bloch, prepared by J. D. Walecka*, World Scientific Publishing Company, Singapore; originally published by Stanford University Press, Stanford, CA (1989)

Walecka, J. D., (2004). *Theoretical Nuclear and Subnuclear Physics, 2nd ed.*, World Scientific Publishing Company, Singapore

Walecka, J. D., (2007). *Introduction to General Relativity*, World Scientific Publishing Company, Singapore

Walecka, J. D., (2008). *Introduction to Modern Physics: Theoretical Foundations*, World Scientific Publishing Company, Singapore

Walecka, J. D., (2010). *Advanced Modern Physics: Theoretical Foundations*, World Scientific Publishing Company, Singapore

Walecka, J. D., (2011). *Introduction to Statistical Mechanics*, World Scientific Publishing Company, Singapore

Walecka, J. D., (2013). *Topics in Modern Physics: Theoretical Foundations*, World Scientific Publishing Company, Singapore

Walecka, J. D., (2017). *Introduction to Statistical Mechanics: Solutions to Problems*, World Scientific Publishing Company, Singapore

Walecka, J. D., (2017a). *Introduction to General Relativity: Solutions to Problems*, World Scientific Publishing Company, Singapore

Walecka, J. D., (2018). *Introduction to Electricity and Magnetism*, World Scientific Publishing Company, Singapore

Walecka, J. D., (2019). *Introduction to Electricity and Magnetism: Solutions to Problems*, World Scientific Publishing Company, Singapore

Walecka, J. D., (2020). *Introduction to Classical Mechanics*, World Scientific Publishing Company, Singapore

Walecka, J. D., (2021). *Introduction to Classical Mechanics: Solutions to Problems*, World Scientific Publishing Company, Singapore

Walecka, J. D., (2021a). *Introduction to Quantum Mechanics*, World Scientific Publishing Company, Singapore

Wentzel, G., (1949). *Quantum Theory of Fields*, Interscience, New York, NY

Wiki (2021). *The Wikipedia*, http://en.wikipedia.org/wiki/(topic)

Index

www.ingramcontent.com/pod-product-compliance
Lightning Source LLC
Chambersburg PA
CBHW071603200326
41519CB00021BB/6846